THE MILITARY GUIDE TO

SPIRITUAL WARFARE

USING MODERN WARFARE TACTICS
TO PREDICT SATAN'S ATTACKS

BY

Shane W. Cunningham

IRON SHIELD
PRESS

San Antonio, Texas

Library of Congress Control Number: 2025918682

Hardcover ISBN: 979-8-9930664-1-7
Softcover ISBN: 979-8-218-77872-9
eBook ISBN: 979-8-9930664-0-0
Audiobook ISBN: 979-8-9930664-2-4

IRON SHIELD
PRESS

Published by Iron Shield Press
An imprint of 7 Benih Ministry
San Antonio, Texas
www.ironshieldpress.com
www.7benihministry.org

Scripture quotations are taken from the
New American Standard Bible® (NASB),
Copyright © 1960, 1971, 1977, 1995, 2020
by The Lockman Foundation.
Used by permission. All rights reserved.
(www.lockman.org)

Printed in the United States of America

DEDICATION

Dedicated to my wife, Jessica – I will always fight for you.

Table of Contents

Preface The Rules of Engagement (How to Use this Book)..i

Phase I The Pre-Conflict Phase (Laying the Groundwork) 1

Chapter 1 The Reality of War (Spiritual Awareness).........3

Chapter 2 Defining Objectives (Our Downfall)................ 13

Chapter 3 The Art of Deception (Tactical Planning) 31

Chapter 4 Coalition Building (The Enemy's Hierarchy)..47

Chapter 5 The Strategy of Deterrence (Preventing Conflict) 63

Chapter 6 Preparation & Mobilization (Positioning His Forces) 77

Phase II The Initial Phase (The Assault) 89

Chapter 7 Gaining Air Superiority (Controlling the Narrative) 91

Chapter 8 Information Warfare (Propaganda & Psychological Operations)................................. 107

Chapter 9 Suppression of Enemy Defenses (Silencing the Truth)..121

Chapter 10 The Initial Offensive (Exploiting Our Weaknesses) 133

Chapter 11 Naval Operations – Sea Control (Infiltrating Our Hearts)... 147

Phase III The Siege 157

Chapter 12 Degrading Enemy Capabilities (Severing the Connection) 161

Chapter 13 Destroying Key Infrastructure (The Foundations of Faith).. 173

Chapter 14 Attrition Warfare (Wearing Us Down) 185

Chapter 15 Securing Key Terrain (Claiming Territory in Our Lives) 201

Chapter 16 Isolating the Enemy (Isolating Us from Christ) 215

Chapter 17 Counterinsurgency (Suppressing Our Will to Resist) 227

Phase IV The Consolidation Phase (Spiritual Slavery) 239

Chapter 18 Establishing Security (Controlling Our Thoughts and Actions) 241

Chapter 19 Disarmament and Demobilization (Stripping Us of Our Spiritual Weapons) 249

Chapter 20 Establishing a New Government (Replacing God's Authority with His Own) 257

Chapter 21 Reconstruction and Development (Rebuilding Our Lives in Satan's Image) 263

Chapter 22 Winning Hearts and Minds (Blinding Us to the Truth) 271

Epilogue The Ultimate Rally Point 279

Appendix Your Spiritual Armory (Key Verses for Memorization) 283

Endnotes 289

Bibliography 303

Glossary of Terms 309

About the Author 315

"Know your enemy and know yourself and you can fight a hundred battles without disaster"

-Sun Tzu, *The Art of War*

So that no advantage would be taken of us by Satan, for we are not ignorant of his schemes."

- 2 Corinthians 2:11

Preface

The Rules of Engagement
(How to Use this Book)

EVERY NATION THAT survives does so because it takes its enemies seriously. Armies study tactics, analyze threats, build defenses, and prepare for both offensive and defensive campaigns. They know that to ignore the presence of hostile forces is not a strategy of peace, it is a blueprint for defeat.

The same is true in the Christian life. The Apostle Paul makes this uncomfortably clear:

"For our struggle is not against flesh and blood, but against the rulers, against the authorities, against the powers of this dark world and against the spiritual forces of wickedness in the heavenly places." – Ephesians 6:12

This verse is not poetic imagery. It is a battlefield briefing. It tells us that the conflict in which we live is not metaphorical, not optional, and not fought on human terms. We face a cunning adversary, one who studies, deceives, and strikes with deliberate intent. His campaigns are not random temptations, but calculated strategies designed to fracture faith, undermine trust in God, and leave believers spiritually disarmed.

And yet, far too many Christians live as though this war does not exist. Some ignore the reality of the enemy out of fear, others because of discomfort, and still others because they have been lulled into complacency by the comforts of the world. But ignoring Satan's strategies is like a nation disbanding its military while hostile armies mass on its borders. The danger does not vanish because we refuse to acknowledge it.

Military commanders know that war is never won by accident. Victories require preparation, intelligence, discipline, and decisive action. To assume that spiritual warfare is any less intentional is a fatal miscalculation. Satan does not need to invent chaos; he thrives on predictability. He studies the habits, weaknesses, and vulnerabilities of each believer, looking for entry points. He refines his methods generation after generation, turning subtle lies into strongholds and distractions into weapons.

The purpose of this book is not to sensationalize the demonic or to fill pages with shadowy speculation. Our aim is to equip believers with the knowledge, awareness, and strategies needed to recognize and resist the schemes of the devil (Ephesians 6:11). Just as military doctrine is built on both offensive and defensive operations, so too the Christian must learn how to defend against temptation while also advancing the kingdom of God with confidence and power.

Throughout these pages, we will explore Satan's campaigns through the lens of modern military tactics – reconnaissance, deception, deterrence, attrition, psychological warfare, and more. Each chapter will pair historic battles with biblical truths, showing how earthly warfare mirrors the unseen conflict raging around us.

But this is not merely a book of warnings. It is also a book of hope and victory. The cross of Christ was the decisive invasion, the D-Day of human history, where Jesus disarmed the powers and authorities and made a public spectacle of them (Colossians 2:15). Our fight is real, but it is not hopeless. Our enemy is dangerous, but he is already defeated. The outcome of this war is not in doubt.

So why study Satan's methods at all? Because though the war is won, the battles remain. Soldiers who march carelessly into combat without armor still fall, even when the final victory is assured. Paul's command to "put on the full armor of God" (Ephesians 6:13) is not advice for the super-spiritual; it is the basic training required of every believer.

How This Book Works: A Briefing

This book is designed as a field manual. Each chapter is built on a consistent structure to provide a comprehensive understanding of spiritual warfare by drawing direct parallels between military strategy and the unseen battles Christians face. By understanding this structure, you will be better equipped to use this guide effectively.

Each chapter is divided into four key sections:

Military Warfare: This section begins each chapter by explaining a specific phase or tactic of military warfare. Using historical examples from ancient and modern conflicts, it provides a solid foundation in the strategic principle being discussed.

Spiritual Warfare: This section draws the parallel between the military concept and the strategies used by Satan. It explores how the enemy employs similar tactics in his campaign against believers, using biblical examples and theological insight to expose his playbook.

Counterintelligence: This is your "how-to" section. It provides specific, actionable strategies that you, the believer, can implement to defend against and disrupt the enemy's plans. This is where strategic knowledge turns into personal application.

Biblical Assault: This section serves as our "after-action report," analyzing specific biblical case studies where the enemy's strategies were engaged. We will dissect how God's people triumphed – or failed – and draw direct lessons on how to wage our own spiritual assaults with wisdom and power.

This book is written as both a manual and a charge. It will not flatter you with illusions of neutrality or peace in enemy territory. Instead, it will train your eyes to see the battlefield, your ears to discern propaganda, and your heart to trust in the victory of your King. You will see connections to yourself and the world because the war is already raging both inside and outside your home, here, and abroad.

The war is real. The enemy is cunning. But Christ has already triumphed. The task before us is not to win a war Christ has already secured – it is to stand firm, fight faithfully, and live victoriously until the day the final trumpet sounds.

Perhaps pray the following prayer before continuing – Satan does not want you making it through this book. He attacked me while I wrote it, and he may do the same as you read it:

Heavenly Father, prepare my heart for the words I am about to read. Guard my heart from Satan's whispers and diversions as his plans are revealed to me. Let my knowledge, wisdom, and understanding grow from these historical case studies and the truths from your Word. Open my heart and my eyes. In Jesus' name I pray, Amen.

Phase I

The Pre-Conflict Phase
(Laying the Groundwork)

BEFORE THE FIRST shot is fired, before the clash of steel or the roar of engines, the most decisive battles of any war are fought in silence. This is the Pre-Conflict Phase, a time of intense but often unseen activity where the groundwork for victory is meticulously laid.[1] It is the realm of spies and strategists, of planners and logisticians. It is here that armies study their enemies, define their objectives, develop their strategies, and mobilize their forces.

In this crucial phase, Satan is not idle. He is a master intelligence officer, meticulously conducting reconnaissance on your life. He is a strategist, defining the objectives for your downfall. He is a diplomat, building his dark coalitions. He is preparing the battlefield and positioning his forces, all with the goal of ensuring his victory is a foregone conclusion before the assault even begins.

The chapters in this section will dissect this unseen war. We will analyze the enemy's methods of strategic assessment, his process for defining objectives, his reliance on deception and division, the structure of his demonic hierarchy, and his campaigns of deterrence and mobilization.

This is the shadow war, and it is here, in the quiet preparation, that the foundations of victory are laid.

Chapter 1

The Reality of War
(Spiritual Awareness)

Military Warfare:
The First Battlefield is Knowledge

LONG BEFORE BULLETS fly or artillery shells explode, war begins with silence. The kind of silence that hums with hidden activity – radios whispering coded messages, analysts hunched over grainy images, scouts creeping through forests under cover of night. Every commander knows: the first battlefield is knowledge.[1] Without effective reconnaissance and intelligence, an army is blind, deaf, and destined for defeat. Strategy becomes guesswork, tactics become reckless, and courage becomes a tragic waste.

Case Study: Operation Desert Storm
(1991) – The Air War Before the Ground War

In the winter of 1991, as the world watched Saddam Hussein's forces dig into the Kuwaiti desert, an invisible war was already unfolding. Coalition aircraft thundered off carriers in the Persian Gulf and bases in Saudi Arabia, not for glory, but for reconnaissance and precision strikes. For **thirty-eight nights,** the skies over Iraq burned with fire. Stealth bombers, unseen by radar, slipped past defenses to cripple communication centers. Satellites hovered overhead, mapping the exact locations of tanks hidden under camouflage nets. Electronic warfare units jammed radio frequencies, turning Iraqi orders into static.[2] Pilots spoke of enemy soldiers waving white flags before the ground assault even began[3] – so confused and cut off that they no longer knew what was happening around them.

When U.S. tanks finally rumbled across the border, what should have been a bloody desert struggle turned into a lightning advance. The Iraqi army was not defeated on the battlefield – it had been undone in the shadows, days and weeks before the ground war began. What the world saw in one hundred hours of ground combat was simply the conclusion of a war that had already been won in the realm of intelligence. The lesson was clear: **wars are decided before they are fought.**

Case Study: World War II – The Secrets of Enigma

In the 1940s, Hitler's armies marched across Europe like an unstoppable tide. German U-boats prowled the Atlantic, sinking Allied ships and starving Britain of supplies. Victory seemed uncertain. But in the quiet halls of Bletchley Park in England, mathematicians and cryptographers waged a different kind of war. Day and night, they bent over chalkboards, twisting symbols, straining to unlock the secrets of the German **Enigma machine** – a device thought impossible to crack.

When the breakthrough came, everything changed. Suddenly, Allied commanders could read German messages.[4] They knew where U-boats lurked. They anticipated troop movements. Convoys were rerouted safely, invasions timed to perfection. Winston Churchill later said that the work at Bletchley Park shortened the war by at least two years.[5] Hitler had tanks, planes, and divisions by the thousands. But the Allies had his playbook, and in war, that is the deadliest weapon of all.

4

Spiritual Warfare: Satan's Reconnaissance

This military principle has a chilling spiritual parallel. For just as generals map the battlefield and study their enemies, Satan does the same with us. He has been in reconnaissance mode for millennia. The war you feel today did not begin when temptation struck. It began long before – in the quiet hours when the enemy studied you, catalogued you, and prepared the strike.

Every predator knows that victory begins with patience. A lion does not roar before the hunt; it crouches in the tall grass, eyes fixed, breath silent, waiting for weakness to reveal itself. When Peter warns us that the devil prowls around like a roaring lion seeking someone to devour (1 Peter 5:8), he is not describing chaos – he is describing calculated predation. Satan is not omniscient, but he is observant. He has watched humanity for thousands of years, gathering intelligence like a master strategist.

- **If you have a short fuse,** he notes it and arranges situations to provoke your temper.

- **If you crave approval,** he knows which flatteries will inflate your pride.

- **If you carry insecurities,** he whispers lies that deepen them until you question your worth.

- **If you've endured trauma,** he reopens wounds at opportune moments to breed bitterness.

- **Even your strengths** are noted; courage can be twisted into recklessness, discernment into suspicion, devotion into legalism.

The enemy doesn't guess – he studies. His attacks are not random flares but surgical strikes, aimed at the weakest point in the wall. This is why his temptations often feel so **tailored.** It's not generic; it's customized. The thought that flashes across your mind, the situation that seems to appear at the worst possible moment – it feels designed because it is. While he studies us with precision, many Christians never study themselves. They remain blind to their own weaknesses, unprepared for the attack that has been planned for months, or even years.

Counterintelligence: Defending Against and Assaulting the Enemy

The best militaries know that reconnaissance is inevitable. Spies will come. Satellites will circle. The question is not whether you are being studied – it is whether you are prepared. This is why every army develops **counterintelligence.** This involves not only protecting one's own secrets but actively disrupting the enemy's ability to gather and use intelligence.

In the lead-up to D-Day, British counterintelligence ran a breathtaking operation. Dozens of double agents fed false information to the Nazis. Inflatable tanks and wooden planes created the illusion of an army preparing to invade Calais instead of Normandy.[6] This didn't just protect Allied soldiers; it saved thousands of lives. Victory in this hidden war depended on one thing: **knowing yourself as well as the enemy knew you.**

For the believer, Paul captured this principle perfectly: "so that we would not be outwitted by Satan; for we are not ignorant of his schemes" (2 Corinthians 2:11). Our counterintelligence begins with self-awareness but must mature into a proactive, offensive posture.

Phase 1: Defensive Counterintelligence (Know Yourself)

First, you must conduct reconnaissance on yourself before the enemy does. Just as generals pour over maps, believers must map their own weaknesses.

Keep a Spiritual Journal (Your Battle Map). Write down when, where, and how temptations strike. Over time, patterns will emerge.

Identify Triggers (Mark the Ambush Sites). Is it fatigue that makes you irritable? Social media that fuels envy? A particular person who stirs compromise? Mark these "ambush sites" so you can avoid or guard them.

Name the Root (Expose the Spy). Behind every temptation lies a root. Pride. Insecurity. Fear. Pain. Ask God in prayer to reveal the root cause. Once the spy is unmasked, its power weakens.

Practice Awareness (Stay on Patrol). Remain alert. Test thoughts. Examine motives. Stay watchful. A soldier who naps on watch endangers the whole unit.

Phase 2: Offensive Counterintelligence (Assault and Disrupt)

Defense alone cannot win a war. You must be prepared not only to endure the first strike – but to fight back with decisive force, disrupting the enemy's operations.

Assault with the Word of God. Scripture is the "sword of the Spirit" (Ephesians 6:17), an offensive weapon. When Satan attacks, you counter with precision: "It is written..." Memorization is not busywork; it is weapons training. When fear rises, Psalm 27:1 should spring to your lips. When shame whispers, Romans 8:1 should fire back.

Assault with Prayer. Prayer is not a white flag; it is calling in fire support. When Nehemiah's workers were threatened, they labored with one hand and held a weapon in the other, but Nehemiah also prayed: "Hear us, our God, for we are despised. Turn their insults back on their own heads" (Nehemiah 4:4). Prayer calls the Commander of Heaven's Armies to strike the enemy lines.

Assault with Praise. In 2 Chronicles 20, King Jehoshaphat faced an overwhelming army and sent singers ahead of the soldiers. As they praised, the Lord set ambushes against the enemy. Praise shifts the battlefield, drowns out lies with truth, and reminds the enemy of what he lost.

Assault with Testimony. Revelation 12:11 states that believers overcome Satan "by the blood of the Lamb and by the word of their testimony." Your testimony is an assault on the enemy's narrative. It declares: "You tried to destroy me, but God rescued me." It exposes his lies and weakens his grip.

Biblical Assault:
After-Action Reports on Reconnaissance

These case studies reveal how critical good intelligence – and the faith to act on it – truly is.

Case Study: The Twelve Spies (Numbers 13-14) –
A Mission Sabotaged by Bad Intelligence

This is the quintessential biblical reconnaissance mission, and a catastrophic failure. Moses sent twelve spies, one from each tribe, into Canaan with clear objectives: assess the land, the people, their cities, and bring back a report. For forty days, they scouted the terrain. The mission itself was a success; they gathered the intelligence. The failure came in how that intelligence was interpreted and reported.

The Two Intel Reports: Ten of the spies returned with a fear-based assessment. They confirmed the land was fruitful but then presented the enemy as invincible giants and themselves as mere "grasshoppers." Their intelligence was filtered through a lens of fear and unbelief. Joshua and Caleb presented a faith-based assessment. They saw the same giants but declared, "Do not be afraid of the people of the land, because we will devour them. Their protection is gone, but the LORD is with us" (Numbers 14:9). They saw the enemy through the lens of God's promise.

The Strategic Consequence: The congregation chose to believe the bad intelligence. The fear report led to a full-scale mutiny against God's leadership. The consequence of rejecting good intel and embracing bad was devastating: an entire generation was condemned to wander and die in the wilderness. The war was lost before it even began, not because the enemy was too strong, but because the people's belief was too weak.

Case Study: David vs. Absalom (2 Samuel 15-17) – Victory Through Counter-Reconnaissance

Years later, King David provided an expert example of how to win an intelligence war. When his son Absalom launched a coup, David was forced to flee Jerusalem. He was outnumbered and outmaneuvered. His victory came not on the battlefield, but through a brilliant counterintelligence operation.

Planting the Double Agent: David instructed his loyal counselor, Hushai the Archite, to feign defection and offer his services to Absalom.[6] Hushai became David's spy in the enemy's command center.

Feeding False Intelligence: Absalom's lead strategist, Ahithophel, gave him brilliant advice: attack David immediately while he is weak and disorganized. This would have won the war for Absalom. But Hushai, the double agent, countered with deliberately bad advice. He appealed to Absalom's ego, suggesting he gather a massive army for a grand, glorious battle, thus giving David precious time to escape and regroup.

The Strategic Consequence: Absalom fatefully chose to believe Hushai's bad intelligence over Ahithophel's good intelligence. This delay was all David needed. He organized his forces, chose the battlefield, and decisively crushed Absalom's army. David won because he controlled the intelligence, manipulated his enemy's perception, and turned the enemy's counsel into a weapon against him.

Closing Charge: Be Watchful, Be Ready, Be Armed

You know the truth: the enemy is watching. He is studying. He is patient. He circles like a lion, not wasting energy but waiting for weakness. This war began long before you noticed the arrows flying. It began the moment the enemy turned his gaze on you. But you are not defenseless.

The lessons from the battlefield are clear. In Desert Storm and with the Enigma machine, victory was achieved because one side knew the enemy better than the enemy knew himself. In the same way, Satan is running reconnaissance on you right now. He has studied your habits, your weaknesses, even your strengths. He waits for moments of fatigue, loneliness, pride, or distraction. And when he strikes, it will not feel random. It will feel tailored.

But you have a greater advantage. You are not blind. God has warned you. Scripture has revealed the schemes. You do not have to be caught off guard. Like David, you can use wisdom to counter the enemy's plans. Like Joshua and Caleb, you can interpret the battlefield through the lens of God's promises, not your fears.

This is your practical call:

Be Watchful. Keep your eyes open. Know your battlefield.

Be Honest. Name your weaknesses before the enemy does. Shine light where Satan wants darkness.

Be Armed. Memorize Scripture like a soldier drills with his rifle. Prepare now.

Be Prayerful. Call in fire support. Prayer shifts battles in ways unseen.

Be Bold. Don't just survive. Testify. Praise. Worship. Strike back at the enemy's lies.

You do not fight alone. The Lord of Hosts – the Commander of Heaven's Armies – goes with you. He has given you the armor of God, the sword of the Spirit, and the shield of faith. The enemy may study you, but he cannot defeat the One who dwells within you, for "Greater is He who is in you than he who is in the world" (1 John 4:4).

Therefore, do not slumber. Do not stumble blindly into the ambushes. Do not underestimate the enemy, but do not overestimate him either. He is cunning, but he is not sovereign. He is powerful, but he is not almighty. You are not a victim in this war; you are a warrior, armed and chosen, fighting from a position of victory because the cross has broken the enemy's ultimate power. Stand watch. Stay alert. Sharpen your sword. And when the ambush comes, don't just hold the line – assault the darkness. For this war is not only about survival; it is about conquest, until the kingdom of God fills every corner of your life.

Shane Cunningham

Chapter 2

Defining Objectives (Our Downfall)

Military Warfare: The Decisive Aim

IN THE UNFORGIVING realm of military conflict, victory hinges on more than just manpower and weaponry. Every successful campaign, every decisive battle, is born from a bedrock of clearly defined objectives.[1] A seasoned commander doesn't commit troops to the field with vague aspirations; they deploy them with a precise understanding of what must be achieved: to seize vital terrain, cripple enemy logistics, neutralize a key asset, or ultimately, compel the adversary to surrender. Without these clearly articulated objectives, the theater of war descends into a chaos of wasted resources, eroded morale, and, inevitably, defeat. Military objectives provide clarity, focus resources, measure progress, and adapt plans.[2]

The importance of clearly defined objectives extends beyond the battlefield itself. It permeates every aspect of military planning and execution, from strategic assessments and resource allocation to tactical maneuvers and logistical support. A well-defined objective provides a compass for decision-making, ensuring that every action contributes to the overall goal. It enables commanders to prioritize efforts, allocate resources efficiently, and measure progress effectively. Moreover, it fosters a sense of unity and purpose among the troops, boosting morale and cohesion.

Case Study: The Attrition of Verdun (1916)

The Battle of Verdun, a grueling clash on the Western Front during World War I, serves as a stark example of the devastating consequences of vague objectives.[3] The German strategy, spearheaded by General Erich von Falkenhayn, was not aimed at territorial gain, but rather at inflicting mass casualties on the French army. Falkenhayn believed that by bleeding the French white at Verdun, Germany could force France to seek a negotiated peace. This vague objective, however, lacked a concrete and achievable goal beyond simply inflicting casualties.

The relentless bombardment and attritional warfare resulted in staggering losses on both sides, with little strategic gain for either.[4] The battle became a symbol of the war's futility, a testament to the horrific cost of conflict waged without clear and attainable objectives. It highlights the danger of pursuing a strategy based solely on attrition, without a clear vision of the desired end state.

Case Study: Operation Rolling Thunder (1965-1968)

Operation Rolling Thunder was the title of a gradual and sustained aerial bombardment campaign conducted by the U.S. 2nd Air Division, U.S. Navy, and Republic of Vietnam Air Force against North Vietnam from 2 March 1965 until 2 November 1968, during the Vietnam War.[5] The operation ultimately proved to be a strategic failure, hampered by a lack of clear objectives and a flawed understanding of the enemy's resolve.

The four objectives of the operation were:

1. Boost the sagging morale of the Saigon regime.

2. To persuade North Vietnam to cease its support for the communist insurgency in South Vietnam without actually taking any ground forces into communist North Vietnam.
3. To destroy North Vietnam's transportation system, industrial base, and air defenses.
4. To halt the flow of men and material into South Vietnam.

These objectives, while seemingly straightforward, were often conflicting and unrealistic.[6] The gradual and incremental nature of the bombing campaign allowed North Vietnam to adapt and mitigate the damage. Furthermore, the political constraints imposed by the U.S. government limited the scope and effectiveness of the operation. The failure to achieve its objectives ultimately contributed to the prolonged and ultimately unsuccessful U.S. involvement in the Vietnam War.

Contrasting Clarity: The Six-Day War (1967)

In stark contrast, the Six-Day War between Israel and its Arab neighbors exemplifies the power of clearly defined objectives. Faced with imminent threats from surrounding nations, Israel launched a preemptive strike with the clear aims of neutralizing enemy air forces, seizing strategic territories (including the Golan Heights, West Bank, and Sinai Peninsula), and securing its borders.[7]

These precise objectives guided every aspect of Israel's military operations. The result was a swift and decisive victory that dramatically altered the geopolitical landscape of the Middle East.[8] The clarity of objectives allowed Israel to concentrate its forces, exploit enemy weaknesses, and achieve its goals with remarkable speed and efficiency.

Case Study: Operation Entebbe (1976) – Precision and Purpose

Operation Entebbe (1976) also shows how key objectives can be essential for victory.[9] This operation's main objective was to rescue 103 mostly Israeli hostages at Entebbe Airport in Uganda.

Israeli commandos flew over 2,500 miles to rescue the hostages held by terrorists backed by the Ugandan army.

The mission was fraught with risk, requiring meticulous planning, precise execution, and unwavering resolve.[10] The Israeli commandos were given a clear and unambiguous objective: to rescue the hostages and bring them home safely. This singular focus guided their actions throughout the operation, enabling them to overcome seemingly insurmountable obstacles and achieve a remarkable success.

The objectives were so clear that the forces were able to maintain discipline in a chaotic environment. Of the 103, 3 hostages died while another 5 terrorists and 45 Ugandan soldiers died. The operation was a huge success which wouldn't have been possible without clear objectives.

The Broader Strategic Context

The importance of clearly defined objectives extends beyond tactical victories; it is essential for achieving broader strategic goals.[11] A nation's military objectives must align with its overall political and economic interests. They must be realistic, achievable, and sustainable. A failure to define and pursue coherent strategic objectives can lead to protracted conflicts, wasted resources, and ultimately, strategic defeat.

Military objectives provide clarity, focus resources, measure progress, and adapt plans. The lessons of military history are clear: victory in warfare requires more than just firepower and courage. It demands a clear understanding of the objectives, a well-defined strategy for achieving them, and the unwavering commitment of all forces involved. Without these essential elements, even the most powerful military can be led astray, squandering its resources and sacrificing its soldiers in pursuit of elusive and ultimately unattainable goals.

Spiritual Warfare: The Enemy's Calculated Assault

Unlike the chaotic clashes of brute force, spiritual warfare is a theater of calculated objectives. Satan does not launch random, undirected assaults against believers. He wages war with defined strategic goals, aiming to undermine our faith, diminish our effectiveness, and ultimately, sever our connection with God. Jesus, in His profound insight, laid bare the enemy's sinister agenda: "The thief comes only to steal and kill and destroy; I have come that they may have life, and have it to the full" (John 10:10).

The specific manifestations of the enemy's objectives may vary from individual to individual, shaped by our unique vulnerabilities and circumstances. However, the underlying destructive DNA remains consistent:

To Steal: To rob us of our joy, our peace, our spiritual effectiveness, and our inheritance in Christ. This can be our relationships (family, friendships), ministry opportunities, even our physical health (through stress, anxiety, and unhealthy habits).

To Kill: To extinguish our testimony, poison our relationships, and stifle our spiritual vitality. This involves more than physical death; it's about killing our passion for God, our desire to serve others, and our ability to make a meaningful impact on the world.

To Destroy: To demolish our intimacy with God, dismantle our ability to advance His Kingdom, and ultimately, secure our eternal separation from Him. This is the ultimate goal, the complete and utter ruin of our lives and our destinies.

Case Study: Samson
(Judges 16) – A Life Dismantled by Design

The tragic downfall of Samson, chronicled in Judges 16, offers a chillingly clear illustration of Satan's defined objectives at work.

The enemy's strategy wasn't simply to inflict momentary humiliation on the mighty judge; it was to systematically dismantle him, stripping him of his strength, his purpose, and his relationship with God. The enemy's objectives were clear:

To Steal: To erode Samson's strength by enticing him to compromise his Nazarite vow, the very source of his supernatural power. This was not a sudden, impulsive act but a gradual erosion of his commitment, fueled by Delilah's persistence and Samson's own growing pride and lust.

To Kill: To neutralize Samson's divinely ordained mission to deliver Israel from the Philistines, rendering him incapable of fulfilling his calling. By compromising his vow, Samson forfeited his ability to lead and protect his people, effectively killing his God-given purpose.

To Destroy: To obliterate Samson's identity as a man of God, severing his connection with the divine and leaving him vulnerable to despair and destruction. This involved not just physical captivity but a complete spiritual and emotional breakdown, leading to his tragic death.

Delilah's relentless persistence was not a mere quirk of personality; it was a carefully orchestrated campaign, fueled by the enemy's unwavering commitment to his destructive objectives.

The Broader Strategic Context: Satan's Long-Term Game

It's crucial to understand that Satan's objectives are not limited to individual sins or isolated incidents. He's playing a long-term game, seeking to establish strongholds in our lives and in the world around us. He systematically targets key areas:

Our Minds: By planting seeds of doubt, fear, and anxiety (2 Timothy 1:7), he seeks to control our thoughts and cloud our judgment. He uses misinformation and propaganda to shape our perception of reality and lead us astray. (2 Thessalonians 2:9) He attacks our minds with negative thoughts, doubts, anxieties, and fears (2 Corinthians 10:5). He seeks to destabilize our emotions and cloud our judgment. He plants seeds of doubt about God's love, power, and faithfulness. He tells us that we're not good enough, that we'll never change, and that God has given up on us.

Our Hearts: He seeks to infiltrate our hearts with pride, selfishness, lust, anger, bitterness, and other destructive emotions (Proverbs 4:23). He wants to turn us away from love and compassion, making us self-centered and insensitive to the needs of others. He may try to convince us that we're entitled to certain things or that we deserve to be treated better than others.

Our Relationships: He uses division to weaken the body of Christ, fostering discord and conflict among believers. (James 3:16) He sows seeds of envy, jealousy, and resentment, creating rifts in our families, churches, and communities.

Our Environments: He subtly shifts the cultural landscape to make it easier for us to fall. He infiltrates our environments (Ephesians 2:1-2), subtly influencing our thoughts, attitudes, and behaviors. He fuels our flesh (1 John 2:16) by providing opportunities to indulge in sinful desires.

He also wants to cause rapid deployment: The enemy attacks us at our weakest points, exploiting our vulnerabilities and past traumas to trigger emotional responses and lead us into sin (James 1:14). He knows our triggers and pushes our buttons. He uses temptation to lure us away from God, promising short-term pleasure at the expense of long-term fulfillment. This is quick and often sudden, giving the initial grip of our lives.

Satan knows that we are stronger together, and that our relationships with other believers provide us with encouragement, support, accountability, and spiritual nourishment. Therefore, he actively seeks to isolate us from the church. The goal is to create attrition warfare in our lives. Attrition warfare focuses on **gradually wearing down an enemy's strength** through sustained combat, aiming to deplete their resources and ability to fight, rather than achieving decisive breakthroughs or strategic maneuvers. Satan's attrition warfare is characterized by a constant, often subtle barrage of temptations, trials, challenges, and discouragements, aiming to deplete our spiritual, emotional, and even physical reserves. He understands that persistent, subtle attacks are often more effective than dramatic, frontal assaults. He doesn't necessarily need to make us fall into a huge sin; he just needs to make us tired enough that we stop resisting, stop caring, and stop actively pursuing God.

By understanding these broader strategic objectives, we can be better equipped to recognize and resist the enemy's attacks, not just in individual moments of temptation but in the overall direction of our lives.

Counterintelligence: Defining and Defending Your Mission

In the theater of war, the most effective counterintelligence is not merely reactive; it is fiercely proactive. It is not enough for an army to simply identify an enemy's positions; it must understand the enemy's ultimate objectives. An intelligence officer doesn't just report "enemy tanks are moving north." They analyze the movement to discern the strategic goal: "The enemy is attempting to encircle our eastern flank to cut off our supply lines." This discernment – the ability to see past the immediate threat to the ultimate objective – is what separates victory from slaughter.

So it is in spiritual warfare. The enemy's objectives have been clearly defined: **to steal, kill, and destroy**. Our counterintelligence, therefore, begins with the crucial discipline of identifying his strategic aims behind every tactical assault he launches against us. We must learn to look past the presenting temptation and ask the critical question: What is the enemy's *endgame* here?

Recognizing the Enemy's True Objectives

Satan's assaults are never isolated events; they are calculated moves designed to achieve a larger, more destructive purpose. The believer who only sees the surface-level temptation is like a soldier who only sees the incoming artillery shell, oblivious to the fact that it is a precursor to a full-scale ground invasion. We must train ourselves to recognize the strategic intent behind the tactical distraction.

When you are tempted with lust, the immediate objective appears to be a moment of impurity. But the strategic goal is far more sinister. It is a targeted strike designed to **destroy intimacy with your spouse,** create a foundation of deceit, and sever your fellowship with a holy God. It is a scorched-earth tactic against the sacred covenant of marriage and your personal walk with Christ.

When bitterness or unforgiveness festers in your heart, the immediate objective seems to be your emotional pain. But the strategic goal is to **establish a fifth column within your soul** and your community. A fifth column is a group of enemies operating secretly from within a defended territory to aid an outside attacker. They are traitors and saboteurs who look like they belong, but their loyalty is to the invading force. An unforgiving spirit gives the devil a legal foothold (Ephesians 4:27), chokes out the fruit of the Spirit, divides relationships, and breaks down the unity of the church – our primary defense network.

When busyness and distraction overwhelm you, the immediate objective appears to be simple exhaustion. But the strategic goal is **deterrence and interdiction.** The enemy aims to keep you so occupied with good things that you have no time for the essential things: prayer, Scripture, and fellowship. He is cutting your spiritual supply lines, ensuring that when a real crisis hits, you are too spiritually malnourished and prayerless to withstand it.

Counterintelligence demands that we move beyond simply resisting sin and begin to dismantle the enemy's strategic objectives. We must interrogate every temptation, every negative thought pattern, every encroaching compromise by asking: *If I give in to this, what territory does the enemy gain? What mission of God in my life is neutralized?*

Establishing Your Commander's Intent: Defining Your Mission

An army without a clear mission is nothing more than an armed mob, wandering aimlessly until it is destroyed. Likewise, a Christian without a clear, God-given purpose is vulnerable to every demonic strategy and cultural whim. The most powerful form of counterintelligence is to operate with such a strong sense of your own mission that the enemy's objectives become irrelevant distractions.

Your life is not a defensive crouch; it is an offensive campaign under the command of Jesus Christ. His objective for you is clear: "I have come that they may have life and have it to the full" (John 10:10). Your personal mission statement must be an extension of this divine Commander's Intent. It is the declaration of what you will achieve for the Kingdom of God, providing the "why" behind every battle you fight. This isn't a vague platitude; it is your operational order, defining your purpose and directing your resources. It might be: "To glorify God by raising my children to be disciples of Christ," or "To be a beacon of integrity and godly counsel in my secular workplace," or "To use my gifts of teaching to equip the saints for the work of ministry."

Without this clarity, you will constantly be reacting to the enemy's initiatives. With it, you force the enemy to react to yours.

Practical Counterintelligence Directives

To counter Satan's objectives, you must be ruthlessly intentional. The following directives will help you defend your territory and stay on mission.

Conduct a Personal Threat Assessment. Be brutally honest with yourself. Where are you most vulnerable? Is it pride after a success? Despair after a failure? A specific person or environment that triggers compromise? Like a military planner studying terrain, you must map your own heart, identifying the weak points in your defenses that the enemy is most likely to exploit. This is the essence of being "sober-minded" and "watchful" (1 Peter 5:8).

Define and Post Your Mission Statement. A mission that isn't written down is just a wish. Take time in prayer to discern God's primary objectives for this season of your life. Write them down. Post them where you will see them daily – on your bathroom mirror, on your desk, as the wallpaper on your phone. This is your constant reminder of what you are fighting for.

Interrogate Every Temptation. When a tempting thought, a destructive emotion, or a compromising opportunity arises, do not be passive. Take that thought captive (2 Corinthians 10:5) and subject it to interrogation. Ask it: "What is your objective? Who sent you? What ground do you seek to take?" By exposing the enemy's goal, you rob the temptation of its deceptive power.

Guard Against Mission Drift. Just as the U.S. war effort in Vietnam was crippled by a lack of clear objectives, a believer's life can be neutralized by mission drift. Busyness, even with good activities or ministry work, can pull you away from your primary objectives of intimacy with God and fulfilling your specific calling. Regularly ask yourself: "Do my daily activities align with my stated mission, or have I been diverted?"

Establish Mission-Focused Accountability. Your Battle Buddies (accountability partners) are more than just sin-confessors; they are your fire team, your unit. Their role is not just to ask, "Did you stumble this week?" but to ask, "Are you staying on mission? Are your actions and priorities aligned with the objectives God has given you?" This type of accountability reinforces your mission and helps you identify and correct mission drift before it becomes catastrophic.

Living without a defined, God-given purpose makes you a soft target. But a believer who understands the enemy's objectives and is fiercely committed to their own is a formidable warrior. You are no longer just a defender of a position; you are an assaulter of darkness, advancing the Kingdom of God with clarity, focus, and divine power. This is how you hold the line. Now, we will learn how to assault the enemy's position.

Biblical Assault:
Demolishing the Enemy's Objectives

To master the art of biblical assault, we must become students of past campaigns. Scripture does not just give us commands; it provides detailed after-action reports from the front lines of spiritual warfare. These are not just stories; they are case studies in how the strategic objectives of the enemy have been met and demolished by the power and wisdom of God. An assault, in this context, is the decisive action taken to neutralize the enemy's core strategy. By dissecting these biblical counter-offensives, we learn to recognize the deeper patterns of spiritual conflict and how to apply God's time-tested principles to the battles we face today.

Case Study: Israel at Sinai
(Exodus 32) – Assaulting a National Betrayal

While Moses was on the mountain receiving the law – the very terms of God's covenant with Israel – a spiritual insurgency was brewing in the camp below. The people, impatient and fearful, compelled Aaron to forge a golden calf.

Satan's Strategic Objective: This was far more than a simple act of idolatry. The objective was to **annul the covenant at its inception.** Satan's goal was to destroy Israel's trust in God's leadership, prove them unworthy of the Promised Land, and provoke a righteous God into destroying them completely. The golden calf was an attempt to trigger a strategic self-destruction of God's chosen people.

The Biblical Assault: The assault came from two directions.

The Assault of Intercession (Moses): Before God's judgment could fall, Moses launched a brilliant counter-offensive in the heavenly realms. He didn't beg for mercy based on the people's merit; he appealed to God's own character and strategic interests. He reminded God of His covenant promises to Abraham, Isaac, and Jacob, and argued that destroying Israel would damage God's reputation among the Egyptians (Exodus 32:11-13). This was high-level intercession, holding God's own Word and reputation up as a shield against judgment.

The Assault of Righteous Action (Moses): Upon returning to the camp, Moses executed a swift and decisive ground assault. He physically destroyed the idol, grinding it to powder and making the Israelites drink it – forcing them to ingest the bitter taste of their sin. This was a physical act that demolished the enemy's symbol of authority and purged the treason from the camp.

Case Study: Peter's Sifting (Luke 22:31–34) – A Preemptive Assault by Christ

On the eve of the crucifixion, Jesus revealed a direct, personal threat against His chief disciple. Peter, full of self-reliant bluster, was completely unaware that he was the target of a specific demonic operation.

Satan's Strategic Objective: As Jesus stated, Satan's goal was to "sift you like wheat." This wasn't merely about causing Peter to stumble in fear. The strategic objective was to **shatter Peter's confidence so completely that he would feel disqualified from any future leadership role.** The goal was to take the future "rock" of the church and grind him into dust, neutralizing him as a leader forever.

The Biblical Assault: The assault here was launched not by Peter, but *for* him, by Jesus Himself.

The Assault of Revelation: Jesus exposed the enemy's plan before it happened. "Satan has demanded..." This act of intelligence sharing immediately armed Peter with awareness, even if he didn't grasp its full importance.

The Assault of Intercession: Jesus then revealed His counter-offensive: "but I have prayed for you, that your faith may not fail." Christ's own prayer was the divine missile defense system that would ensure Peter's core faith survived the attack, even if his courage faltered.

The Assault of Re-Commissioning: Critically, Jesus gave Peter his new mission *before* the failure: "and when you have turned back, strengthen your brothers." Jesus assaulted Satan's objective (disqualification) by pre-authorizing Peter's restoration and reframing his coming failure as a prerequisite for a future ministry.

Case Study: Ananias & Sapphira (Acts 5) – Assaulting a Fifth Column

The early church was an outpost of heaven on earth, characterized by radical unity, purity, and power. This made it a prime target for infiltration. Ananias and Sapphira sold a piece of property and lied about the amount they gave, pretending a piety they did not possess.

Satan's Strategic Objective: This was not simply about greed. The objective was to **corrupt the purity of the early church at its foundation.** It was an attempt to establish a "fifth column" – a secret cell of hypocrisy and deception – within the body of Christ, normalizing lies and quenching the unique power of the Holy Spirit that rested on their authenticity.

The Biblical Assault: The assault was a swift, Spirit-led surgical strike executed by the Apostle Peter.

The Assault of Discernment: Peter, filled with the Holy Spirit, did not see two people telling a white lie. He saw the enemy's strategic objective. He bypassed the human actors and addressed the spiritual source of the attack: "Ananias, how is it that Satan has so filled your heart that you have lied to the Holy Spirit?" (Acts 5:3).

The Assault of Divine Judgment: The immediate and severe judgment that fell on Ananias and Sapphira was a divine counter-offensive. It violently purged the enemy's fifth column from the camp. This act served as a terrifying warning, establishing a "no-fly zone" for deception and protecting the integrity of the early church's mission. The result was that "great fear seized the whole church" (Acts 5:11), reinforcing the high stakes of their holy calling.

From these campaigns, we learn that a biblical assault is not blind aggression, but a targeted, discerning counter move against the enemy's true strategic goals, waged through intercession, truth, and decisive, Spirit-led action.

Closing Charge: Define and Conquer

You have been presented with the intelligence. The battlefield of your life is not a scene of random skirmishes, but a theater of war defined by competing objectives. An enemy strategist has meticulously studied you, identified your vulnerabilities, and defined clear, destructive goals for your life. His campaign is already in motion. The time for passive observation is over. The time for decisive action is now.

The enemy's objectives are brutally clear, a dark reflection of John 10:10. He seeks to **steal** your joy and your peace through the slow attrition of doubt and discouragement. He aims to **kill** your effectiveness by cutting you off from your purpose and causing you to drift into a comfortable, lukewarm faith. His ultimate goal is to **destroy** your intimacy with God, leaving you feeling isolated and defeated. This is his unwavering mission.

But you do not serve a commander of chaos. You do not fight under a banner of vague intentions. Your Commander-in-Chief, Jesus Christ, has issued His own objective for your life, and it is the polar opposite of the enemy's: "I have come that they may have life, and have it to the full." This is not a suggestion; it is God's strategic goal for you. His objective is not your mere survival, but your fruitfulness. His plan is not your containment, but your advancement. While the enemy plots your ruin, God has decreed your restoration and your victory.

Therefore, your charge is to counter the enemy's objectives by being ruthlessly committed to God's. The following are not mere recommendations; they are your operational directives for engaging the enemy and fulfilling your mission.

Define Your Mission in Christ. Your first order is to receive your Commander's Intent. An army that doesn't know what it's fighting for will fall apart under fire. Prayerfully seek God and write down your personal mission statement. What has God called you to do? "To love God, serve my family, and disciple others." "To be a person of integrity and a light in my workplace." This is your operational order. Let it govern every decision. A clear mission is the foundation of all victorious warfare.

Discern the Enemy's Goals. You must become a student of your enemy's strategy. When temptation strikes, do not see it as a random event. See it as a tactical move with a strategic purpose. Ask: *If I yield to this, what does Satan truly gain?* If you give in to discouragement, he steals your momentum. If you give in to distraction, he kills your focus. By unmasking the ultimate objective, you strip the immediate temptation of its power.

Refocus on God's Objectives. When the fog of war descends – when you are weary, confused, and under fire – you must have rally points. Your primary rally point is the objective of your King. Remember that His goal for you is not just to avoid sin, but to bear much fruit. His objective is not just to keep you safe, but to make you dangerous to the kingdom of darkness. In moments of trial, shift your focus from the enemy's attack to your Commander's promise. This is how you reclaim the high ground.

Guard Against Mission Drift. A believer's life can be neutralized by mission drift. The enemy will gladly let you be busy with a dozen good things if it keeps you from the one essential thing God has called you to do. Busyness that chokes out prayer is mission drift. Ministry devoid of love is mission drift. Relentlessly evaluate your life: Do my actions, my schedule, and my resources align with my defined mission, or have I been subtly diverted?

Establish Mission Accountability. No one fights alone. You need a team – a battle buddy, a mentor, a small group – who knows your mission. Their job is not just to ask if you've been good, but to ask if you've been on mission. "Are you pursuing the objectives God gave you? Where are you being diverted? How can we help you get back on track?" This is the iron-sharpening-iron that keeps you effective in the fight.

What campaign will your life be? Will it be a long, draining, and ultimately fruitless conflict with no clear objective? Or will it be a life marked by clear purpose, decisive action, and glorious victory for your King? The enemy has objectives for your life. It is time you made it clear you have objectives of your own – objectives handed down from the throne of heaven. Know your mission. Discern the enemy's plans. Demolish his objectives. Execute your Commander's will.

Shane Cunningham

30

Chapter 3

The Art of Deception
(Tactical Planning)

Military Warfare: The Strategy of Illusion

ONCE A COMMANDER has defined the objectives of a campaign, the next critical phase begins: developing a strategy.[1] This is the intellectual heart of warfare, the comprehensive plan that dictates *how* objectives will be achieved. A strategy is far more than a simple battle plan; it is a complex architecture of coordinated efforts. It involves selecting which battlefields to engage on and which to avoid, allocating finite resources for maximum impact, timing assaults to exploit enemy vulnerabilities, and, most critically, employing the sophisticated art of deception. An army that marches openly and honestly toward its objective is an army that will be met, countered, and annihilated.

"All warfare is based on deception."[2] This ancient principle, famously articulated by the Chinese general Sun Tzu over two millennia ago, remains the immutable bedrock of military strategy. It is the ghost that haunts every battlefield, the unseen force multiplier that can transform a weaker force into a victor and a stronger force into an invincible one. Deception is not a minor tactic or a clever trick; it is a core strategic principle designed to attack the most critical and vulnerable enemy asset: the commander's mind. By manipulating an enemy's perception of reality, a skilled strategist can make them misinterpret strengths as weaknesses, see threats where none exist, and commit their best forces to defending the wrong ground.

The methods of military deception are as varied as the conflicts they shape. They range from simple tactical feints on the battlefield – a small unit attacking on one flank to draw defenders away from the real, main assault on the other – to grand, strategic deceptions that shape the course of entire wars. Camouflage, the most basic form of deception, hides troop movements and conceals critical assets from aerial and satellite reconnaissance. Disinformation, spread through covert channels and intelligence operatives, can sow confusion and mistrust within an enemy's command structure, leading to paralysis and poor decision-making. In the modern era, electronic warfare can create phantom armies on radar screens, while cyber operations can disable communication networks or feed false intelligence directly into an enemy's systems.

Underpinning all of this is the art of psychological operations (PSYOPs). These are campaigns designed to erode enemy morale, encourage desertion, and create a sense of hopelessness that can cripple the will to fight long before the first shot is fired.

Case Study: Hannibal at the Battle of Cannae (216 B.C.)

Long before the advent of complex intelligence networks, the Carthaginian general Hannibal Barca demonstrated on the plains of Cannae that the most devastating weapon of war is the mind of a brilliant strategist. Facing a Roman army that vastly outnumbered his own, Hannibal knew a direct, force-on-force engagement would mean certain annihilation. His entire strategy, therefore, was built on a masterful deception designed to turn

Rome's greatest strength – its powerful, aggressive infantry charge – into its fatal weakness.

Hannibal arranged his forces in a highly unconventional formation.[3] He placed his weakest infantry, the Gauls and Spanish, in the center of his line and pushed them forward in a crescent shape, bulging toward the Romans. He held his elite, veteran African infantry back on the flanks, almost hidden from the main line of sight.

As the battle began, the Romans did exactly what Hannibal predicted. Their powerful legions charged forward, crashing into the weak Carthaginian center. The Gauls and Spanish began to give way, bending backward under the immense pressure. To the Roman commanders, it looked like a rout; they believed they were breaking the enemy line and pushed their reserves forward to exploit the "victory." But it was a trap. The Carthaginian line was not breaking; it was bending like a bow, luring the entire Roman army deeper and deeper into a pocket.

Once the Roman legions were fully committed, packed shoulder-to-shoulder and unable to maneuver effectively, Hannibal sprang his trap. His elite African infantry, who had been waiting patiently on the flanks, wheeled inward and slammed into the exposed sides of the Roman formation. Simultaneously, Hannibal's superior cavalry, having defeated their Roman counterparts, circled around and attacked the Roman rear. The mighty Roman army was completely encircled. What followed was not a battle, but a slaughter. Hannibal had used a brilliantly deceptive strategy to achieve one of the most decisive tactical victories in military history, proving that a superior plan can defeat a superior force.

Case Study: Operation Fortitude (1944) – The Ghost Army

Perhaps the most breathtaking and successful strategic deception in modern history was Operation Fortitude, the Allied plan to deceive the Germans about the location of the D-Day landings. As the Allies massed forces in Britain for the 1944 invasion of Normandy, they knew that success hinged on one critical factor: preventing the Germans from concentrating their elite Panzer divisions at the landing sites. To achieve this, they orchestrated a massive, multi-layered deception campaign designed to convince the German High Command that the main invasion would occur at the Pas-de-Calais, the narrowest point of the English Channel.

The Allies created an entire phantom army, the First U.S. Army Group (FUSAG), "stationed" in southeast England, directly opposite Calais.[4] This ghost army was given all the signatures of a real fighting force. It had inflatable tanks, dummy landing craft, and plywood aircraft parked at deserted airfields to fool enemy reconnaissance planes. It had a famous and feared commander, General George S. Patton, whose reputation for aggression made him the logical choice to lead the main assault. It generated enormous amounts of fake radio traffic, mimicking the communications of a massive force preparing to strike.

Crucially, a network of double agents, controlled by British intelligence, fed a steady stream of "credible" information to their German handlers, all confirming that the main thrust would come at Calais. The plan worked with devastating effectiveness. Even after the landings at Normandy began on June 6, 1944, Hitler and his generals remained convinced that Normandy was merely a diversionary attack.[5] He held his most powerful Panzer divisions in reserve near Calais for weeks, waiting for the "real" invasion that would never come. That critical hesitation, born of a masterfully executed deception, gave the Allies the precious time they needed to establish a secure beachhead and begin the liberation of Europe. Operation Fortitude is the ultimate proof that the most powerful weapon in a general's arsenal is the ability to control what the enemy believes to be true.

Spiritual Warfare: The Father of Lies

If deception is a key strategy in human warfare, it is the *native language* of our spiritual adversary. Satan is not an honest broker of temptation; he is a master of illusion, a purveyor of half-truths, and an architect of division. His entire kingdom is built upon a framework of deception. He is, as Jesus identified him with chilling accuracy, "a murderer from the beginning, not holding to the truth, for there is no truth in him. When he lies, he speaks his native language, for he is a liar and the father of lies" (John 8:44). This is his default setting, his core nature. He does not need to overpower us with irresistible force if he can first mislead us with a compelling lie.

His primary strategic objective has never changed: to create a rift between humanity and God. To achieve this, he employs a devastatingly effective two-pronged strategy: first, to deceive the mind, and second, to divide the fellowship.

Phase 1: The Strategy of Deception

Satan's initial assault is always one of information warfare. He rarely presents sin as the ugly, destructive force it truly is. Instead, he packages it in appealing disguises, wrapping poison in the veneer of wisdom, freedom, or justice. His method is not to create falsehoods out of thin air, but to take a legitimate truth and twist it just enough to lead us astray.

This playbook was established in his very first encounter with humanity. In Eden, his attack on Eve was a masterful tactical deception. He didn't begin with a command, but with a subtle, insidious question designed to undermine her confidence in God's clarity and goodness: *"Did God really say...?"* (Genesis 3:1). With that single question, he introduced doubt. He then escalated his assault by impugning God's motive, suggesting God was a celestial tyrant withholding something good. Finally, he issued a direct contradiction to God's command: "You will not certainly die."

This three-step process – questioning the truth, questioning the motive, and deny the consequences – remains his signature move. He used it against Jesus in the wilderness, quoting Scripture but twisting its context to tempt Christ toward pride (Matthew 4:6).

He uses it against us today, whispering lies that sound plausible because they are rooted in our own insecurities and desires:

o "You're not good enough for this ministry; you should just quit." (A lie targeting insecurity).

o "God doesn't really care about that 'small' sin; His grace covers it anyway." (A lie that twists grace into a license for compromise).

o "You're better off on your own; those people in the church will just hurt you." (A lie designed to lead to the second phase of his strategy).

Phase 2: The Tactical Goal of Division

Deception is the seed; division is the poisonous fruit. A lie, once believed, inevitably leads to a breakdown in unity. If Satan can successfully deceive us, he can then divide us – from God and from each other. A divided army cannot stand. Its lines of communication are broken. Its supply lines are disrupted. Its soldiers become isolated, confused, and vulnerable. Satan knows that if he cannot destroy the church from the outside through persecution, he can neutralize it from the inside through division. As James warns, "For where you have envy and selfish ambition, there you find disorder and every evil practice" (James 3:16).

He seeks to create factions, to magnify minor offenses into major conflicts, and to turn brothers and sisters in Christ into adversaries. In Corinth, he divided the church over allegiance to charismatic leaders (1 Corinthians 1:12). In Galatia, he used the deceptive doctrine of legalism to create a deep rift between Jewish and Gentile believers. His goal is always the same: to get us to "bite and devour one another" so that we are "destroyed by one another" (Galatians 5:15), saving him the effort.

Biblical Case Study: The Insurgency of Korah (Numbers 16)

Nowhere is this strategy of deception leading to division more violently illustrated than in the rebellion of Korah. Korah, a prominent Levite, became resentful of the authority of Moses and

Aaron. He launched a spiritual insurgency, and his primary weapon was a masterful deception.

He approached Moses not with an outright lie, but with a dangerous half-truth: "You have gone too far! The whole community is holy, every one of them, and the LORD is with them. Why then do you set yourselves above the LORD's assembly?" (Numbers 16:3). This was a lie wrapped in the language of pious humility.[7] Was the community holy? Yes. Was the Lord with them? Yes. But Korah twisted this truth to launch a direct assault on God's ordained leadership structure.

His deception was wildly successful. He fueled division by appealing to the pride and ambition of other leaders, gathering 250 "well-known community leaders" to his side. He had successfully split the camp, creating a direct mutiny against God's chosen commanders. The result was a catastrophic divine judgment. The earth literally opened up and swallowed Korah and his co-conspirators, and fire from the Lord consumed the 250 leaders who had joined him. This terrifying event underscores the severity with which God views division born from deceptive, prideful ambition. It demonstrates that an attack on God-ordained unity is an attack on God Himself.

Satan's strategy is timeless. He will seek to deceive you with subtle questions and appealing lies. He will tempt you to doubt God's goodness and to question His Word. And if you allow that deception to take root, his next move will be to use it to divide you from the very people who form your spiritual defensive line. He doesn't need to storm the fortress if he can convince the soldiers inside to turn on each other.

Counterintelligence:
The War for Truth and Unity

To counter an enemy whose primary strategy is deception and whose primary tactic is division, our counterintelligence must be relentlessly focused on two fronts: the front of absolute truth and the front of unbreakable unity. In modern intelligence work, this is known as protecting against infiltration and subversion. An agency not only verifies external intelligence to protect against enemy deception, but it also vets its own people and reinforces internal protocols to protect against spies and saboteurs who would sow discord from within. A lie from the outside is a threat, but a lie believed on the inside is a catastrophe.

Our counterintelligence, therefore, cannot be a passive defense. It must be a set of active, disciplined measures designed to create an environment in our minds and in our communities where lies cannot survive and division cannot take root. The enemy's entire campaign hinges on our acceptance of falsehoods. We will counter by becoming experts in the truth and guardians of our fellowship.

Directive 1: Vetting All Intelligence (Testing the Spirits)

No intelligence officer would ever act on a single, unverified report from an unknown source. Every piece of intel is scrutinized, cross-referenced, and vetted before it is considered actionable. This is a matter of operational survival. The Apostle John issues the same standing order to every believer: "Do not believe every spirit, but test the spirits to see whether they are from God, because many false prophets have gone out into the world" (1 John 4:1).

This is our primary security protocol. Every teaching, every prophecy, every "new idea," every book, every podcast, and every influential personality must be subjected to rigorous testing. The litmus test is singular and absolute: the Word of God.

o   Does this teaching align with the known character of God as revealed in the entirety of Scripture?
o   Does it elevate Jesus Christ and affirm the sufficiency of His work on the cross?
o   Does it produce the fruit of the Spirit – love, joy, peace, patience – or does it produce pride, division, and confusion? (Galatians 5:22-23).

Testing the spirits is not an act of cynicism; it is an act of supreme discernment. It is the spiritual equivalent of a customs checkpoint, refusing to allow any counterfeit doctrine or demonic deception to infiltrate the borders of our minds and churches.

Directive 2: Mastering Your Doctrine (Knowing the Whole Counsel)

To effectively test incoming intelligence against a known standard, you must first be an absolute master of that standard. An agent trained to spot counterfeit currency doesn't study every possible fake; they study the real thing so intensely that any flaw in a forgery becomes immediately and glaringly obvious. Likewise, to become deception-proof, we must immerse ourselves in the "whole counsel of God" (Acts 20:27).

Satan is an expert at quoting Scripture selectively. His temptation of Jesus in the wilderness is the prime example. He quoted from Psalm 91, inviting Jesus to throw Himself from the temple. It was a direct quote, but it was twisted to serve a prideful, deceptive purpose. Jesus countered the assault not by arguing the specific verse, but by deploying a deeper, contextual truth from Deuteronomy: "It is also written: 'Do not put the Lord your God to the test'" (Matthew 4:7). He knew the *whole* doctrine, not just isolated proof-texts.

A casual, surface-level familiarity with the Bible is insufficient – it actually makes you *more* vulnerable to sophisticated deception. The enemy preys on biblical illiteracy. Our counterintelligence requires a deep, abiding, contextual knowledge of the Word of God. This is what transforms us from being potential victims of deception into being active detectors of it.

Directive 3: Fortifying Internal Cohesion (Guarding Unity)

A commander knows that an enemy's psychological operations are most effective against a unit with low morale and internal strife. Division is a self-inflicted wound that the enemy is more than happy to exploit. If Satan's tactical goal is to divide us, then our counter-tactic must be to fiercely and intentionally guard our unity.

Unity is not a passive state of simply "not fighting." It is an active, strategic pursuit. It requires constant maintenance.

Humility: Recognizing that we are not always right and that the mission of the Body is more important than our personal agenda.

Forgiveness: Obeying the command to not let the sun go down on our anger (Ephesians 4:26), because unresolved anger is a "foothold" for the devil – an open door for a demonic saboteur to enter our camp.

Grace: Extending to others the same grace that has been extended to us, refusing to hold onto offenses or participate in gossip that tears down a fellow believer.

Every time you choose to forgive, every time you refuse to repeat a negative story, every time you believe the best in a brother or sister, you are engaging in an act of counter-warfare. You are repairing a potential breach in the wall, reinforcing your unit's cohesion, and neutralizing the enemy's strategy of division.

Directive 4: Active Counter-PSYOPs
(Identifying Subtle Lies)

Finally, our counterintelligence must be personal. The main battlefield for deception is the territory between our ears. Satan's psychological operations are the subtle, whispered lies that flash through our minds, disguised as our own thoughts:

o   "God doesn't really care about you; if He did, this wouldn't be happening."
o   "You're better off alone. You can't trust anyone in the church."
o   "Your sin isn't that bad. It's not like you're hurting anyone."
o   "You've failed too many times. You'll never change."

Our directive here is clear: "We demolish arguments and every pretension that sets itself up against the knowledge of God, and we **take captive every thought to make it obedient to Christ** " (2 Corinthians 10:5). This is an aggressive, proactive command. When a lie-based thought enters your mind, you are to arrest it. Challenge its credentials. Interrogate it against the truth of Scripture. Find it guilty of being an enemy agent and execute it by replacing it with the truth. When the lie says, "You're better off alone," you assault it with the truth: "Let us not give up meeting together... but let us encourage one another" (Hebrews 10:25).

This mental discipline is our internal counter-propaganda campaign. By refusing to give sanctuary to the enemy's lies in our minds, we starve his entire strategy of the fuel it needs to operate. An individual committed to truth and a community committed to unity is a fortress that cannot be overthrown by deception.

Biblical Assault: A Study of Counter-Deception

The war for truth is not new, and Scripture serves as our combat history, filled with detailed accounts of the enemy's primary strategy of deception and his tactical goal of division. By analyzing these campaigns, we learn to recognize his methods, expose his vulnerabilities, and execute our own assaults against his ancient and unchanging strategy.

Case Study: The Gibeonite Deception (Joshua 9)

After the stunning victories at Jericho and Ai, the Israelite army was a force to be feared, operating with divine momentum. The enemy, recognizing that a direct military confrontation was likely futile, shifted to a strategy of deception. The Gibeonites, a local Canaanite tribe, knew they were marked for destruction and orchestrated a masterful psychological operation to save themselves.

The Deception: They did not come with weapons, but with a fabricated story. They staged a scene, arriving with worn-out clothes, cracked wineskins, and dry, moldy bread. Their entire appearance was designed to sell a single lie: "We have come from a distant country." This lie was critical, as Israel was forbidden from making treaties with the inhabitants of Canaan but could make them with distant nations. It was a perfectly crafted deception, designed to bypass Israel's rules of engagement.

The Failure of Counterintelligence: The leaders of Israel fell for the deception completely. The critical failure is noted in verse 14: "The Israelites sampled their provisions but **did not inquire of the LORD.** " They relied on their own senses – what they saw and heard – instead of consulting their Commander for divine intelligence. They vetted the enemy's story with their eyes, not with the Spirit. This failure to seek divine counsel was the breach the enemy needed.

The Resulting Division and Consequence: Joshua and the leaders made a treaty, swearing an oath before God. When the deception was discovered three days later, it caused a massive internal problem. The entire assembly "grumbled against the leaders" (v. 18), creating division and undermining the leadership's authority. Because of their oath, Israel could not destroy the Gibeonites, and this deceptive infiltration created a strategic and spiritual complication that would trouble Israel for generations.

Case Study: The Threat of Internal Strife (Acts 6)

In the early days of the Church, when its unity and power were a direct threat to the kingdom of darkness, the enemy attempted to sabotage the mission from within.

The Deception (Implied): The enemy's strategy was to exploit a legitimate logistical issue – the daily distribution of food – and weaponize it. A complaint arose from the Greek-speaking Jews that their widows were being overlooked. This was a subtle but dangerous deception, planting the idea of favoritism and injustice within the leadership. The goal was to fracture the church along pre-existing cultural lines, distracting the apostles and bogging the ministry down in internal conflict.

The Biblical Assault: The apostles' response was a brilliant counter-offensive. They did not get defensive. They identified the enemy's true objective: to distract them from their primary mission of "prayer and the ministry of the word." Their assault was a strategic reorganization. By appointing seven Spirit-filled men to oversee the distribution, they decisively neutralized the source of division, empowered new leaders, and protected their own strategic focus. The result was that "the word of God spread" (Acts 6:7). They countered a strategy of division with a strategy of wise delegation and renewed focus on the mission.

The Commander's Warning (Galatians 5:15)

The Apostle Paul, a veteran of countless spiritual campaigns, issued a stark warning to the churches in Galatia being torn apart by deceptive doctrine. His warning reads like a grim battlefield assessment: "If you bite and devour one another, watch out or you will be destroyed by one another." This is not the language of simple disagreement. It is the language of mutual annihilation. Paul understood that internal division accomplishes the enemy's objective for him, as the church consumes its own strength and testimony in bitter infighting.

The historical record is clear: deception is the gateway to disobedience, and division is the weapon that weakens God's people from within.

Closing Charge: The War for Truth

have seen the enemy's foundational strategy, the ancient and unchanging doctrine upon which his entire kingdom is built: deception. His goal is to attack your mind, to twist your perception of reality until you can no longer distinguish the truth from the lie. He is a master of illusion, an architect of false narratives. Like the ghost army of Operation Fortitude, he will construct entire phantom realities – fears that don't exist, accusations that aren't true, promises that are empty – hoping you will exhaust your spiritual and emotional energy fighting shadows.

From this core strategy of deception flows his most effective tactical outcome: division. A lie, once believed, inevitably creates a rift. A lie about God will divide you from your source of strength. A lie about yourself will divide you from your God-given identity. And a lie about a brother or sister in Christ will divide you from your fellowship, leaving you isolated and vulnerable. He knows he does not need to launch a frontal assault against a church that is busy destroying itself from within.

But you have not been left defenseless in this information war. You have been given the ultimate counter-deception weapon: the absolute, unchanging truth of God. Your charge, therefore, is not to live in fear of the lie, but to become a warrior for the truth, creating an environment in your life where deception cannot survive and division cannot take root.

These are your operational directives:

Immerse Yourself in the Truth. Your primary defense is saturation. The Word of God is the only antidote to the enemy's poison. Daily immersion in Scripture is not a religious chore; it is the process of calibrating your mind to reality. The more you know the real thing, the more instantly you will recognize the counterfeit. Make the truth of God's Word more real and more authoritative to you than your own feelings or the whispers of the enemy.

Pray for Discernment. Ask your Commander for divine intelligence. Pray for the Holy Spirit to give you a sanctified skepticism – the ability to look at a teaching, a thought, or a situation and sense when something "sounds right but isn't." This is the spiritual gift of seeing through the enemy's camouflage and identifying the trap before it is sprung.

Guard Your Words. Every believer is a communications officer. You can either transmit the truth of God, which builds up, or you can become an unwitting agent in the enemy's propaganda campaign by spreading gossip, criticism, and negativity. Refuse to be a mouthpiece for division. Commit to speaking life and edification, building a culture of honor that starves the spirit of division.

Seek Reconciliation Quickly. An unresolved conflict is an open gate in your city wall, an invitation for the enemy to enter and establish a stronghold. Do not give him that opportunity. Obey the command to not let the sun go down on your anger (Ephesians 4:26-27). Be the first to apologize. Be quick to forgive. A humble and reconciled heart is a fortress that the enemy cannot breach.

Stay Accountable in Community. The Gibeonites succeeded because the leaders of Israel acted alone, failing to inquire of the Lord. An isolated believer is the enemy's favorite target. Commit to authentic, honest fellowship. Allow others to speak into your life, to test your thinking, and to hold you accountable to the truth. A cord of three strands is not easily broken.

Affirm the Shared Mission. Division thrives when personal agendas outweigh the corporate mission. Constantly remind yourself and others what you are fighting *for*. You are fighting for the glory of God and the advancement of His Kingdom. When the mission is clear and central, petty disagreements and personal ambitions are exposed for the trivial distractions they truly are.

You are at a crossroads. You can allow your mind to become a battlefield of confusion, leading to a life of broken relationships and spiritual isolation. Or you can make your mind a fortress of truth, and your life a beacon of unity.

Live in the light. Walk in the truth. Stand together. And you will find that the enemy's most sophisticated deceptions are powerless against a life surrendered to the God who *is* Truth.

Chapter 4

Coalition Building
(The Enemy's Hierarchy)

Military Warfare: The Power of Alliances

IN THE GRAND history of human conflict, victory has rarely been the prize of a solitary nation. From the ancient world to the modern age, the formation of coalitions and alliances has consistently been a decisive factor, a force multiplier that can overwhelm even the most formidable individual power.[1] A coalition is more than a simple aggregation of armies; it is a complex strategic entity that multiplies strength, broadens the resource base, provides political and moral legitimacy, and presents an enemy with a multi-front problem that is exponentially more difficult to solve than a single-front war. A nation fighting alone can be isolated and defeated; a nation fighting as part of a united coalition becomes part of a strategic goliath.

The logic of coalition warfare is undeniable. It pools economic and industrial power, allowing for the sustained production of war materiel on a scale no single nation could manage. It combines diverse military expertise and technologies, creating a more versatile and adaptable fighting force. Strategically, it allows for the opening of multiple fronts, compelling an enemy to divide its forces, stretch its supply lines, and defend a vastly larger territory. A sole nation might have a superior army, but it can be bled dry by a thousand cuts from a coordinated alliance.

Case Study: The Allied Powers in World War II

The Second World War stands as the ultimate testament to the power of coalition warfare. In the early stages of the conflict, Hitler's Germany, with its technologically advanced Wehrmacht and innovative Blitzkrieg tactics, appeared unstoppable. It conquered Poland, swept through France, and dominated the European continent. However, Germany and its Axis partners ultimately faced a global coalition of overwhelming power.[2]

The alliance of the United States, the United Kingdom, and the Soviet Union – despite their deep ideological differences – created a strategic nightmare for the Axis.[3] The industrial might of the United States became the "arsenal of democracy," supplying not only its own forces but those of its allies. The British Empire provided critical naval power, intelligence assets (like the codebreakers at Bletchley Park), and an unsinkable island base from which to launch the air war and the eventual invasion of Europe. On the Eastern Front, the Soviet Union absorbed the main weight of the German army, grinding down millions of Hitler's best troops in a brutal war of attrition.

This grand coalition meant Germany could never concentrate its full strength on one objective. While fighting a massive land war in the East, it had to defend against a strategic bombing campaign from the West, fight a naval war in the Atlantic, and counter partisan movements across occupied Europe. The Allied victory was not the triumph of one nation, but the inevitable outcome of a successfully coordinated global coalition that brought more resources, more manpower, and more industrial capacity to bear than the Axis could possibly withstand.

Case Study: The First Gulf War (1991)

A more modern example of a swift and successful coalition can be found in the response to Iraq's 1990 invasion of Kuwait. The United States, while possessing overwhelming military power, understood that acting alone would be politically fraught and strategically less effective. Instead, it spearheaded the formation of one of the largest multinational coalitions in history.

Operation Desert Storm was not a solely American operation; it was the work of a 35-nation coalition that included not only NATO allies like the United Kingdom and France, but numerous Arab states such as Saudi Arabia, Egypt, and Syria. This broad coalition achieved several key strategic objectives before the first shot was fired.[4]

Political Legitimacy: The participation of Arab nations completely undermined Saddam Hussein's attempt to frame the conflict as an imperialist war of the West against the Arab world. It isolated him politically and morally.

Strategic Basing: The cooperation of Saudi Arabia and other Gulf states provided the necessary land and air bases from which to launch the campaign, a logistical necessity that would have been impossible otherwise.

Resource Sharing: Coalition partners contributed financially, militarily, and logistically, sharing the immense burden of the operation.

When the war began, Iraq faced not one enemy, but a united world. The coalition presented a seamless, technologically superior, and politically unassailable front that Saddam Hussein could not hope to overcome. The rapid victory was as much a triumph of diplomacy and coalition building as it was of military might. These historical examples underscore a timeless truth: even the most powerful armies recognize the strategic necessity of fighting alongside allies. No wise commander chooses to fight alone if they can stand with a coalition.

Spiritual Warfare:
The Kingdom of Darkness

Just as no earthly empire goes to war alone, our adversary does not operate as a solitary agent. To believe we are fighting a single, isolated demon is to dangerously underestimate the enemy. Satan commands a vast, disciplined, and ancient coalition, a structured kingdom with a clear chain of command and a unified objective: to oppose the work of God and to bring humanity under its dominion. The Apostle Paul, in his strategic briefing to the church at Ephesus, provides the critical intelligence on the nature of this enemy force. He pulls back the curtain to reveal that our conflict is not a simple skirmish against flesh and blood.

"For our struggle is not against flesh and blood, but against the rulers, against the authorities, against the powers of this dark world and against the spiritual forces of evil in the heavenly realms." – Ephesians 6:12

This single verse is one of the most important pieces of enemy intelligence in the entire Bible. It is not a poetic metaphor; it is a doctrinal org chart of the kingdom of darkness. It reveals that the enemy we face is an organized, hierarchical, and multi-layered coalition.[5] Let's break down this intelligence:

Against the Rulers (ἀρχάς, archas)

This refers to the highest echelon of the demonic command structure.[6] These are the master strategists, the princes of darkness who likely govern vast spiritual territories and craft the long-term campaigns against nations, cultures, and the global Church. They are the enemy's Joint Chiefs of Staff, operating in the unseen realms to direct the course of the war.

Against the Authorities (ἐξουσίας, exousias)

This implies delegated power.[7] These are the field commanders, the demonic governors and generals who execute the strategies developed by the rulers. They wield significant authority within their assigned sectors, carrying out the orders of their superiors and commanding the forces under them.

Against the Powers of this Dark World (κοσμοκράτορας, kosmokratoras)

Literally "world-rulers", this points to spiritual forces that exert immense influence over the worldly systems we inhabit.[8] They are the unseen manipulators behind corrupt political ideologies, anti-Christian philosophies, cultural decay, and global systems that promote greed, godlessness, and oppression. They work to shape the very environment in which the spiritual war is fought.

Against the Spiritual Forces of Evil (πνευματικά, pneumatika)

This represents the rank-and-file of the demonic army.[9] These are the countless evil spirits, the foot soldiers assigned to carry out specific tactical missions against individuals, families, and churches.

This is not a disorganized mob. It is a structured kingdom, a demonic coalition operating with chilling efficiency and a shared, hateful purpose.

Case Study: The Gerasene Demoniac (Mark 5)

The raw power and numerical strength of this coalition are put on stark display in the encounter between Jesus and the Gerasene demoniac. When Jesus confronted the tormented man, He did not address a single entity. He addressed a command structure. His question, "What is your name?" was a demand for identification. The reply is one of the most terrifying in all of Scripture: "My name is Legion, for we are many" (Mark 5:9).

A Roman legion was the primary fighting unit of the most powerful army on earth, comprising three to six thousand highly disciplined soldiers. This was not a random name; it was a declaration of military occupation. This man was not merely oppressed; his soul was an occupied territory, garrisoned by an entire demonic battalion. This reveals several critical pieces of intelligence:

They Operate in Coordinated Numbers: The enemy can and does assign a massive number of forces to a single target if the objective is deemed important enough.

They Have Unity of Purpose: The "legion" spoke with one voice and acted with one will – to torment their host and, later, to beg Jesus not to send them out of the region.

They Have a Command Structure: The term "Legion" implies a commander, organization, and discipline. This was not chaos; it was a coordinated military unit on a mission.

Case Study: The Prince of Persia (Daniel 10)

While the story of Legion reveals the enemy's tactical numbers, the prophet Daniel gives us a rare glimpse into the strategic, high-level command. When Daniel prayed and fasted for 21 days, the angel Gabriel was dispatched immediately with the answer. However, the angel was physically resisted and delayed for three weeks. By whom?

Gabriel explains: "But the prince of the Persian kingdom resisted me twenty-one days. Then Michael, one of the chief princes, came to help me" (Daniel 10:13). This "prince of Persia" was not a human king. It was a high-ranking demonic ruler – an *archa* – assigned to the Persian Empire with the strategic objective of influencing its policies and hindering the purposes of God concerning Israel. This encounter reveals that major geopolitical events on earth are mirrored by massive spiritual conflicts in the heavenly realms. It took the intervention of Michael, an archangel and a "chief prince" in God's angelic army, to break through the blockade. This confirms the reality of a powerful, territorial demonic hierarchy actively working to oppose the will of God on a global scale.

This is the coalition we face. It is organized, with specialists in lust, fear, addiction, and deception. It is coordinated, layering temptations to create a cascade of failure – fear feeding into doubt, doubt feeding into sin, and sin feeding into shame. It is strategic, with forces assigned to individuals, families, churches, and entire nations. To stand against such a formidable and united kingdom of darkness while fighting alone is not just foolish; it is strategic suicide.

Counterintelligence: Building the Unbreakable Coalition

In the face of a coalition war, the primary objective of any counterintelligence agency is twofold: to understand the enemy's command structure and to disrupt it, while simultaneously ensuring the strength and integrity of one's own alliances. A commander who allows his forces to be isolated, cut off from communication, and picked apart one by one has already lost the war, regardless of his own army's strength. The enemy coalition, as revealed in Ephesians 6, is vast, organized, and unified in its purpose. Therefore, our counterintelligence cannot be a solitary endeavor. It must be a corporate discipline.

Satan's primary strategy in this theater of war is **isolation.** He knows he is facing the global Church, the Body of Christ – a divine coalition empowered by the Holy Spirit. He cannot defeat this entire force head-on. Therefore, his goal is to break individual units away from the main army. He whispers the strategic lie that you are better off alone, that you don't need the church, that fellowship is optional, and that accountability is a sign of weakness. He seeks to transform a soldier in an army into a lone wanderer in the wilderness. Because an isolated believer is a vulnerable believer – an easy target for the coordinated assault of a demonic legion.

Our counterintelligence, then, is the deliberate, strategic, and relentless pursuit of a holy coalition. We must recognize that our strength is not merely in our vertical relationship with God, but in our horizontal relationships with the other members of His army.

Directive 1: Commit to Your Primary Fighting Formation (The Local Church)

The local church is not a weekly social club or a resource for spiritual goods and services. In the context of spiritual warfare, it is your primary fighting formation. It is your barracks, your command center, your supply depot, and your field hospital all in one. It is the place where you are equipped through teaching, strengthened through worship, and deployed for service. To neglect the local church is the equivalent of a soldier deserting his unit to try and fight the war by himself. It is strategic madness.

Therefore, our first counterintelligence directive is to move beyond casual attendance to active, committed participation. It means showing up for duty, not just when it is convenient or entertaining. It means serving, giving, and investing in the health of the unit. It means submitting to the leadership God has put in place and contributing to the overall mission. A thriving local church is a fortress, a well-supplied garrison that the enemy cannot easily overrun.

Directive 2: Establish Elite Fire Teams (Accountability Partnerships)

Within the larger army, the most effective fighting is often done by small, elite units – the fire team, the squad, the battle buddies who know each other's strengths and weaknesses implicitly. This is the role of an accountability partnership. This is a holy coalition of two or three individuals who are committed to radical honesty, vulnerability, and mutual defense.

This is where you share your specific intelligence; the areas where you are being tempted, the lies you are battling, the fears that are creeping in. An enemy that operates in the darkness of secrecy is defeated when exposed to the light of confession (1 John 1:7). An accountability partner is the fellow soldier in your foxhole who keeps watch when you are weary, who challenges you when you are drifting toward compromise, and who speaks the truth of God's Word when you are starting to believe the enemy's propaganda. Pursuing this level of relationship is not a sign of weakness; it is a mark of a wise warrior who refuses to fight alone.

Directive 3: Coordinate Fire Support (Prayer Networks)

A solitary prayer is powerful. A corporate prayer is devastating to the enemy. Jesus promised a unique spiritual force multiplier when believers form a prayer coalition: "Again, truly I tell you that if two of you on earth agree about anything they ask for, it will be done for them by my Father in heaven. For where two or three gather in my name, there am I with them" (Matthew 18:19-20).

This is our spiritual fire support coordination center. When we form prayer partnerships and networks, we are combining our faith to target specific enemy strongholds. We are moving beyond just praying for our own needs and are beginning to wage offensive warfare on behalf of our brothers and sisters, our church, and our community. We must be intentional about creating these prayer coalitions, understanding that our unified voices call down a level of heavenly intervention that the enemy's hierarchy cannot withstand.

Directive 4: Leverage All Assets (Intergenerational Wisdom)

A successful coalition leverages the unique strengths of all its partners. A wise army does not send its new recruits into battle without the guidance of seasoned veterans, nor does it dismiss the energy and fresh perspective of the young. Our holy coalition must be generational.

The enemy loves to create division between the young and the old. He tells the young that the elders are irrelevant and out of touch. He tells the elders that the youth are reckless and foolish. Our counterintelligence is to intentionally bridge this gap. Young believers desperately need the wisdom, stability, and life experience of the elders who have fought these battles for decades. And the elders need the passion, zeal, and energy of the youth to keep the army advancing. Mentoring and being mentored is not just a discipleship program; it is a strategic imperative for ensuring the long-term health and fighting effectiveness of the entire Body of Christ. As Ecclesiastes reminds us, "Two are better than one... If either of them falls down, one can help the other up. But pity anyone who falls and has no one to help them up" (Ecclesiastes 4:9-10).

Biblical Assault:
After-Action Reports on Coalition Warfare

The principle of coalition warfare is not merely a human strategy; it is a divine one, woven into the fabric of God's redemptive plan. Scripture is filled with accounts of how God's people, when united in a holy coalition, achieved victory, and when isolated, faced defeat. These are not just stories of friendship; they are strategic case studies demonstrating that our strength is magnified and God's kingdom is advanced through unified, collaborative effort.

Case Study: The Battle for Rephidim
(Exodus 17) – Tactical Coalition Support

While Israel was camped at Rephidim, they came under a full-scale assault from the Amalekites. This was their first major military test after leaving Egypt. Joshua was appointed to lead the army on the ground, but the key to victory was not in the valley; it was on the hill above. Moses stood overlooking the battle, holding up the staff of God. As long as his hands were raised, Israel prevailed. When his hands grew tired and fell, the Amalekites gained the upper hand.

The battle hinged on a single, critical point of failure: Moses's physical endurance. Alone, he would have failed, and the army would have been slaughtered. But he was not alone. The Bible reports the decisive counter-maneuver: "When Moses' hands grew tired, they took a stone, put it under him, and he sat on it. Aaron and Hur held his hands up – one on one side, one on the other – so that his hands remained steady till sunset" (Exodus 17:12).

This was a perfect, small-unit coalition. Aaron and Hur identified the strategic weak point – their commander's fatigue – and moved to reinforce it. They did not take over command or try to fight the battle themselves. They supported their leader, allowing him to fulfill his unique role. The result was a decisive victory for Israel. This is a powerful demonstration of tactical support: a small, committed team providing the strength needed at the most critical moment to ensure the success of the entire operation.

Case Study: The Covenant of David and Jonathan (1 Samuel 18-20) – A Strategic Alliance

The relationship between David and Jonathan was more than a deep friendship; it was a strategic alliance that proved critical to David's survival and his eventual ascension to the throne. Jonathan, as the son of King Saul and heir apparent, should have viewed David as a rival. But instead, "he loved him as he loved himself" (1 Samuel 18:3), and they formed a covenant.

This covenantal coalition thwarted King Saul's repeated attempts to assassinate David, who was operating as a lone agent hunted by the state.

Intelligence Sharing: Jonathan acted as David's high-level informant, providing him with critical intelligence about Saul's murderous intentions, allowing David to evade capture multiple times.

Logistical Support: Jonathan provided David with his own weapons and armor, equipping him for survival.

Moral Fortification: When David was at his lowest, a fugitive in the wilderness, Jonathan sought him out "and helped him find strength in God" (1 Samuel 23:16). He reinforced David's resolve and reminded him of God's promise.

This alliance demonstrates that a coalition of two, bound by a covenant and a shared purpose in God, can successfully counter the machinations of a corrupt king and the entire apparatus of a state. It is a testament to the power of a faithful ally in the face of overwhelming opposition.

Case Study: The Early Church
(Acts 2:42-47) – The Ultimate Holy Coalition

The early church, born at Pentecost, provides the ultimate model of a holy coalition in action. Their explosive growth and spiritual power were a direct result of their radical commitment to unity and fellowship. They were not a loose collection of individual believers; they were a single, unified body, operating with a shared heart and mind.

The book of Acts provides their operational rhythm: "They devoted themselves to the apostles' teaching and to fellowship, to the breaking of bread and to prayer" (Acts 2:42). This four-part strategy created an unstoppable spiritual movement.

Unified Doctrine: Devotion to the apostles' teaching ensured they were all operating from the same playbook, protecting them from the enemy's attempts to divide them with false doctrine.

Radical Fellowship: They shared everything they had, selling property to give to anyone in need. This eliminated the internal strife that comes from greed and class division.

Constant Prayer: Their corporate prayer life brought down the power of God, fueling miracles and emboldening their witness.

Their unity was so powerful and attractive that they "enjoyed the favor of all the people," and "the Lord added to their number daily those who were being saved" (Acts 2:47). The early church demonstrates that when God's people operate as a true, Spirit-filled coalition, they become an irresistible force for the Kingdom of God.

Case Study: The Missionary Fire Team
(Acts 13) – Paul and Barnabas

When the Holy Spirit decided to launch the first major missionary offensive to the Gentile world, He did not send a single agent. The Spirit said, "Set apart for me Barnabas and Saul for the work to which I have called them" (Acts 13:2). They were sent out together, as a team. This model of partnered ministry is a divine strategy. Throughout their journeys, they provided mutual encouragement in the face of persecution, accountability in their teaching, and complementary gifts in their ministry. This partnership allowed them to plant churches, confront demonic opposition, and spread the gospel across nations in a way that would have been nearly impossible for a lone individual. It establishes the principle that critical kingdom assignments are meant for coalitions, not lone wolves.

God's Kingdom is advanced by coalitions of faith. These after-action reports prove that from the battlefield to the palace, from the megachurch to the missionary road, victory is found in unity.

Closing Charge: Stand Together

The intelligence is undeniable: you are not fighting a lone wolf; you are facing a kingdom. The Apostle Paul provided the reconnaissance, mapping the enemy's command structure: rulers, authorities, and powers of this dark world. It is a disciplined hierarchy, a demonic coalition operating with a single, malevolent will. It is a Legion, "for they are many," united in its objective to oppose the work of God and to isolate His people. This is the formidable, unified force arrayed against you.

But God, the master strategist, never intended for you to face this coalition alone. His design for His people has always been one of corporate strength and unbreakable fellowship. Therefore, the enemy's primary strategy against you is not a frontal assault, but one of **isolation.** He will whisper the lie that you are stronger on your own, that the church is full of hypocrites, that accountability is suffocating, and that you can manage your spiritual life in private. He seeks to sever your supply lines, cut your communications, and lure you away from your unit, because an isolated believer is a defeated believer.

Do not fall for his oldest trick. Your charge is to actively and intentionally build and maintain the holy coalition God has provided for you. This is not optional; it is a matter of strategic survival and spiritual victory.

These are your operational directives:

Commit to Your Garrison. Your local church is not a building; it is your garrison, your forward operating base in this spiritual war. Do not be a civilian contractor who visits when it's convenient. Be a committed member of the unit. Participate in the fellowship. Serve in the ministry. Submit to the leadership. A strong, healthy church is a fortress that the enemy dreads to attack.

Forge Your Fire Team. Seek out and establish accountability partnerships. This is your fire team, the two or three individuals who are in the foxhole with you. These are the people who know your real struggles, who have permission to ask the hard questions, and who will cover you in prayer when you are under fire. Refuse to live in the isolation of hidden sin.

Call in Combined Fire Support. A solitary prayer is a rifle shot; a praying church is an artillery barrage. Make it your mission to join with others in prayer. The enemy's hierarchy trembles when a coalition of believers agrees together on earth, for they know it unleashes the power of heaven.

Leverage Every Asset. The Body of Christ is an intergenerational army. Seek out the wisdom of the seasoned veterans who have fought these battles for decades. At the same time, invest in and empower the passionate new recruits who bring fresh energy to the fight. A church that honors both its elders and its youth is a powerful and enduring force.

Guard the Unity of the Formation. Your primary mission is greater than your personal preference. Reject the enemy's attempts to sow division through gossip, criticism, and offense. Be quick to forgive and eager to reconcile. The unity of the Body of Christ is a direct assault against the kingdom of darkness.

Will you allow the enemy to isolate you, to pick you off as a straggler far from the main force? Or will you anchor yourself in the strength of God's people? Will you be the hands that hold up your leaders, like Aaron and Hur held up the arms of Moses? Will you be the friend who strengthens another in God, like Jonathan did for David?

The enemy's kingdom is united in its hate. Let the Body of Christ be more united in its love and its mission.

Stand **together**.

Chapter 5

The Strategy of Deterrence (Preventing Conflict)

Military Warfare:
Winning Without Fighting

THE ULTIMATE VICTORY in warfare is not to crush an enemy in a bloody battle, but to dissuade them from ever fighting at all.[1] This is the sophisticated art of deterrence. It is a pre-conflict strategy designed to shape an adversary's decision-making process, convincing them that the potential costs of aggression far outweigh any possible gains. A nation that masters deterrence wins wars before they begin, not by force of arms, but by the credible *threat* of overwhelming force. It is a psychological campaign waged in the mind of the enemy, where the primary objective is to make the very idea of conflict seem irrational and self-destructive.

Military deterrence operates on a foundation of calculated strength and clear communication. It is not enough to simply possess power; that power must be visible, understood, and believed by the potential adversary. This is achieved through several key mechanisms. A **demonstration of capability** –

through large-scale military exercises, naval fleet movements, or showcasing advanced weaponry – sends an unambiguous signal of readiness and strength. **Strategic alliances,** like the NATO pact, serve as a powerful deterrent by creating a collective security agreement; an attack on one member is an attack on all, dramatically raising the stakes for any aggressor.[2] Finally, **clear communication of "red lines"** through diplomatic channels and public statements ensures there is no miscalculation. The goal is to remove all ambiguity, making it plain that certain actions will trigger a swift, certain, and devastating response.

Case Study: The Cold War and Mutually Assured Destruction

For nearly half a century, the United States and the Soviet Union, two superpowers armed with thousands of nuclear weapons, stood on the brink of a conflict that could have ended civilization.[3] Yet, no direct, large-scale war ever erupted between them. This "long peace," as it has been called, was maintained by the most terrifyingly effective deterrent strategy ever conceived: Mutually Assured Destruction, or MAD.

The logic of MAD was as simple as it was horrifying. Both superpowers developed a "second-strike capability," meaning that even if one nation launched a surprise nuclear attack, the other would still have enough surviving weapons – in hardened silos, on submarines, and in long-range bombers – to launch a retaliatory strike that would completely annihilate the attacker.[4] There was no winning a nuclear war; there was only mutual obliteration.

This reality created a state of perpetual, high-stakes deterrence. The cost of initiating a direct conflict was not just defeat, but the end of your own nation. This deterrent was so powerful that it prevented countless potential flashpoints, from the Cuban Missile Crisis to standoffs in Berlin, from escalating into a full-blown war. The Cold War was not "won" in a traditional sense on a battlefield; it was a conflict managed and contained for decades through the sheer, paralyzing power of a credible deterrent.[5] It is the ultimate example of preventing a war by making the consequences of starting one unthinkable.

Case Study: The Pax Romana (Roman Peace)

In the ancient world, the Roman Empire achieved a remarkable period of prolonged peace and stability known as the *Pax Romana,* which lasted for roughly two hundred years.[6] This peace was not born of universal goodwill; it was born of overwhelming deterrent power. While the empire was often engaged in expansion or skirmishes on its distant frontiers, the core territories enjoyed unprecedented security. This peace was maintained by the omnipresent threat of the Roman legions.

The Roman military was not just a fighting force; it was a symbol of absolute power and ruthless efficiency. Any province that considered rebellion, any barbarian tribe that considered a raid, and any rival king that considered an invasion had to weigh their chances against the might of the legions. They knew that challenging Rome would not result in a simple border dispute; it would result in a swift, brutal, and systematic response. The legions were masters of siege warfare, disciplined combat, and punitive retribution. Roman justice was famously merciless to its enemies.

This reputation served as a powerful psychological deterrent. The mere presence of a legionary fortress near a frontier was often enough to pacify an entire region.[7] Potential enemies were deterred not because they fought and lost, but because they calculated the probable cost of conflict and found it to be total annihilation. The *Pax Romana* was a peace kept by the sword – or more accurately, by the fear of the sword ever being drawn. It demonstrates that a reputation for overwhelming and decisive force can be the most effective peace-keeping tool of all.

Spiritual Warfare:
The Preemptive Campaign

Just as military superpowers seek to win wars without firing a shot, our adversary's most effective strategy is often not a direct assault, but a sophisticated campaign of deterrence. His primary objective is frequently not to defeat us in a spiritual battle, but to prevent us from ever entering the command center – the place of intimate communion with God – in the first place. He knows that a believer who is consistently in the presence of their Commander, receiving fresh intelligence, clear orders, and spiritual reinforcement, is a formidable and dangerous foe. Therefore, he wages a relentless, preemptive war to keep us from ever establishing that connection.

Satan's deterrence strategy is designed to make a genuine encounter with God seem unnecessary, unappealing, or impossible. He wants to convince us that we are too busy, too sinful, or too insignificant to approach the throne of God. This is a psychological operation aimed at our will, our emotions, and our priorities. He seeks to create a spiritual "no-man's-land" around the presence of God, using two primary weapons systems: the noise of distraction and the poison of shame.

Weapon System 1: The Noise of Distraction

The modern world is the enemy's perfect battlefield for this tactic. He seeks to deter us from communion with God by filling our lives with an overwhelming amount of static and noise, drowning out the still, small voice of the Holy Spirit. He wants to create a state of perpetual busyness and distraction that leaves no room for quiet reflection or focused prayer. As Jesus warned, the seed of the Word can be "choked by life's worries, riches and pleasures, and they do not mature" (Luke 8:14).

This is not a chaotic, random assault; it is a calculated strategy to degrade our spiritual communications.

Through Busyness: He overloads our schedules with work, social commitments, family obligations, and even ministry activities, convincing us that we are too busy for the one relationship that sustains all others. We become so focused on working *for* God that we have no time to be *with* God.

Through Entertainment: He provides an endless buffet of addictive entertainment – social media feeds, streaming services, video games, televised sports – designed to numb our minds and consume our time. These distractions are not always overtly evil, but they are effective detergents, subtly washing away our desire for the sacred.

Through Worldly Pursuits: He glorifies the chase for wealth, status, and personal achievement, making these idols seem more urgent and rewarding than seeking the Kingdom of God. Our spiritual passion is slowly choked out by our ambition.

The result of this constant noise is **spiritual apathy.** We may still believe in God, but we lose our passion and our devotion. Our faith becomes a background activity rather than the central operating principle of our lives. The enemy has successfully "jammed our communications," not by destroying the tower, but by filling the airwaves with so much static that we no longer even try to listen for the signal.

Weapon System 2: The Poison of Shame

When distraction fails, the enemy deploys his second weapon: shame. He seeks to deter us from God's presence by convincing us that we are unworthy to be there. This is a direct assault on the grace of God. After we sin, he acts as the prosecuting attorney, whispering lies designed to keep us from the very place of forgiveness and restoration.

He blinds the minds of believers to the full reality of the gospel, veiling the truth of God's unconditional love and grace (2 Corinthians 4:3-4). He magnifies our failure and minimizes the cross, whispering insidious lies:

- "How can you go to God after what you just did? You should be ashamed."
- "He's tired of forgiving you for the same thing over and over."
- "You need to clean yourself up first before you can pray."

This is the ancient lie from the Garden of Eden. After Adam and Eve sinned, their first instinct was not to run *to* God for mercy, but to hide *from* Him in shame. Satan's goal is to replicate that dynamic in our lives daily, making us believe that our sin is a barrier to God's presence, when in reality, God's presence is the only solution to our sin. He deters a wounded soldier from entering the field hospital by convincing him he's too bloody to be seen by the doctor.

Case Study: The Deterrence of Martha (Luke 10)

The story of Mary and Martha provides a perfect case study of this strategy. Jesus, the Commander-in-Chief, was physically present in their home. Martha loved Him and was actively serving Him. Her work was good and necessary. Yet, the Bible says she was "distracted by all the preparations" (Luke 10:40).

The enemy used a good thing – service – as a weapon of deterrence. It kept Martha so busy and "worried and upset about many things" hat she was prevented from doing the "one thing [that] is needed" (v.41): sitting at the feet of Jesus, listening to His teaching. Mary chose the better part, the direct communion with her Lord, which would not be taken from her. Martha, deterred by the "noise" of her responsibilities, almost missed the encounter entirely and was left with anxiety and frustration instead of peace. This is a masterful example of how the enemy can use even our service for God to deter us from intimacy *with* God.

This is Satan's preemptive war. He knows if he can keep you out of the throne room through distraction and shame, he may never have to face you on the battlefield. He wins by preventing you from ever receiving your orders.

Counterintelligence:
Breaching the Blockade

In military strategy, countering a deterrence campaign is a high-stakes endeavor. When an enemy has established a credible threat designed to prevent you from acting, the counter move is not to retreat, but to demonstrate that their deterrent is ineffective. It requires a deliberate, calculated decision to breach their blockade, reject their psychological operations, and reestablish contact with allies and command structures. If the enemy's goal is to keep you out of the fight by making you fear the cost of entry, your counterintelligence must be to prove that the cost of *not* acting, of *not* communicating with your Commander, is infinitely higher.

Satan's deterrence strategy against the believer is built on a two-front blockade designed to prevent us from encountering God. He erects a wall of **Distraction** (the noise of the world) and deploys a campaign of **Shame** (the poison of unworthiness). Our counterintelligence, therefore, must be a two-pronged assault designed to breach these specific defenses and open a secure, daily line of communication with our High Command.

Directive 1: Breaching the Blockade of Noise (Countering Distraction)

The enemy has filled the modern world with an unprecedented level of strategic noise. The constant barrage of information, entertainment, and obligation is his electronic warfare, designed to jam our spiritual frequencies and make communion with God seem impossible. To counter this, we must become masters of creating sacred silence and intentionally clearing the airwaves.

Establish a Secure, Non-Negotiable Comms Window: The most crucial step is to schedule a daily, non-negotiable appointment with God. This is not "finding time"; it is *making time*. Treat it with the same seriousness as a strategic briefing with a general. Whether it is 15 minutes before dawn or 30 minutes during a lunch break, this is your dedicated time to open the channel through prayer and Scripture. It is a deliberate act of defiance against the tyranny of the urgent. Guard this time fiercely.

Practice Digital Discipline: Our phones and devices are often the primary delivery systems for the enemy's distracting noise. You must take control of this front. Turn off non-essential notifications. Set specific times for checking email and social media, rather than allowing them to dictate your attention all day. Consider creating "no-phone zones" or times in your home to starve the beast of distraction. This is the equivalent of enforcing radio silence to listen for a critical transmission.

Master the Strategic "No": The enemy will deter you from God's presence by overloading you with good things. Your schedule will be filled with work, family duties, social events, and even ministry activities. You must learn the strategic power of saying "no." Not every opportunity is a divine assignment. If an activity, even a good one, consistently prevents you from maintaining your connection with God, it may be a form of enemy deterrence. Pray for discernment to know which commitments to accept and which to decline for the sake of your primary mission.

Directive 2: Neutralizing Psychological Operations (Countering Shame)

Satan's second deterrent is a psychological operation of shame. He is the "accuser of our brothers and sisters" (Revelation 12:10), and his goal is to make you feel too dirty, too sinful, and too much of a failure to approach a holy God. This is his most ancient and effective lie. Our counterintelligence is to reject his propaganda by relentlessly applying the truth of the Gospel.

Embrace Your Diplomatic Immunity: As a child of God, you have been given what amounts to diplomatic immunity, secured by the blood of Christ. You do not approach God based on your performance but based on your position in His Son. When the enemy accuses you of sin, you do not argue your innocence. You agree with the facts – yes, you failed – but you reject his conclusion. Your counter argument is the cross. You must preach the Gospel to yourself daily, reminding yourself that there is "now no condemnation for those who are in Christ Jesus" (Romans 8:1).

Practice Immediate Confession as a Weapon: In the enemy's economy, your sin is a reason to hide from God. In God's economy, your sin is the very reason you must run *to* Him. Do not allow sin to fester. The moment you are aware of it, confess it immediately (1 John 1:9). This is not an act of groveling; it is an act of war. It neutralizes the enemy's weapon of shame before he can even wield it effectively. You are taking his ammunition and turning it into an opportunity for experiencing God's grace.

Approach the Throne with Confidence: The Bible gives us a direct command that obliterates Satan's shame-based deterrence: "Let us then approach God's throne of grace with confidence, so that we may receive mercy and find grace to help us in our time of need" (Hebrews 4:16). The enemy tells you to approach with fear; God commands you to approach with confidence. This confidence is not in yourself, but in the finished work of Jesus Christ, who has made the way for you. To obey this command is to walk straight through the enemy's psychological barricade and mock his power.

Do not be deterred. The enemy's blockade is an illusion, his psychological warfare is built on lies, and his power is broken. Your access to the throne room has been permanently secured. Your charge is to breach his lines daily and report for duty before your King.

Biblical Assault: After-Action Reports on Breaching the Blockade

The enemy's strategy of deterrence – preventing us from encountering God through distraction and shame – is not a modern invention. It is an ancient campaign. Scripture, however, provides a clear record of victorious counter-offensives. These case studies reveal how focused, determined individuals breached the enemy's blockade and chose the presence of their Commander over the psychological warfare of the world.

Case Study: Nehemiah and the Summit of Ono (Nehemiah 6)

While Nehemiah was engaged in the critical mission of rebuilding Jerusalem's walls, the enemy launched a sophisticated deterrence campaign. Having failed with mockery and threats, Sanballat and his allies shifted to a strategy of distraction. They sent a message to Nehemiah: "Come, let us meet together in one of the villages on the plain of Ono" (Nehemiah 6:2).

The Enemy's Deterrent: This was not a genuine peace summit; it was a strategic diversion. The enemy's objective was to deter Nehemiah from his primary mission. By luring him away from the worksite, they could halt progress on the wall, break his momentum, and potentially assassinate him far from his guards. The invitation was a baited trap, designed to make him abandon his post for a seemingly important diplomatic meeting.

The Biblical Assault: Nehemiah's response was a swift and decisive assault on the enemy's entire strategy. He immediately identified the true objective behind the distraction and refused to be deterred. He sent back a legendary reply: "I am carrying on a great project and cannot go down. Why should the work stop while I leave it and go down to you?" (Nehemiah 6:3). This was not just a refusal; it was a declaration of war on the enemy's strategy. Nehemiah demonstrated perfect mission focus. He refused to allow a secondary, distracting "opportunity" to pull him away from his primary, God-given objective. He assaulted the deterrent of distraction with the weapon of unwavering focus.

Case Study: Mary and Martha (Luke 10)

This encounter provides a powerful side-by-side comparison of succumbing to a deterrent versus breaching it. Jesus was in the house, offering the rare privilege of His direct presence and teaching. The enemy, in this moment, deployed a subtle deterrent using the "noise" of good and necessary service.

The Enemy's Deterrent: Martha was "distracted by all the preparations." Her service was an act of love, but the enemy weaponized it, turning her focus from her Lord to her tasks. The deterrent was the feeling of overwhelming responsibility, the anxiety of hospitality, the noise of a busy home. This distraction successfully kept her from the feet of Jesus, leaving her "worried and upset."

The Biblical Assault: Mary, in contrast, launched a successful assault on the deterrent. She made a conscious, strategic choice. She ignored the noise, bypassed the distractions, and established a position at the feet of her Commander. She chose intimacy over activity. Jesus Himself affirmed her successful breach of the blockade: "Mary has chosen what is better, and it will not be taken away from her" (Luke 10:42). Her assault was a simple, powerful act of prioritization, demonstrating that the most effective way to counter the deterrent of distraction is to deliberately choose the "one thing" that is truly needed: communion with God.

Case Study: The Prodigal Son (Luke 15)

This parable is the ultimate case study in breaching the deterrent of shame. After squandering his inheritance in "wild living," the son found himself destitute, disgraced, and feeding pigs. At this moment, he was the prime target for the enemy's psychological operation of unworthiness.

The Enemy's Deterrent: The lies of the enemy would have been deafening: "You are a complete failure. You have disgraced your family. Your father will never take you back. You are too filthy, too sinful, too far gone. Do not even think of returning; you are not worthy." The objective of this shame-based deterrent is to convince the sinner that the path to repentance is closed, that their sin has created a permanent barrier to the father's presence.

The Biblical Assault: The son's assault began with a moment of clarity: "he came to his senses" (Luke 15:17). He rejected the lie of hopelessness. His assault was his decision to act, even in his unworthiness. He prepared a speech acknowledging his sin and his lack of worthiness ("Father, I have sinned against heaven and against you. I am no longer worthy to be called your son..."). He did not try to clean himself up first. He began the long walk home, covered in the filth of his failure, planning to appeal not for sonship, but for servitude.

The Father's response was the divine counter-offensive that demolished the entire concept of a shame barrier. "But while he was still a long way off, his father saw him and was filled with compassion for him; he ran to his son, threw his arms around him and kissed him" (Luke 15:20). The father did not wait for the apology. He ran to meet his son, closing the distance and obliterating the shame with preemptive grace. The son's assault was his willing return; the father's was his radical acceptance. Together, they demonstrate the most powerful way to defeat the deterrent of shame: to turn toward God, regardless of our perceived worthiness, and to trust in His character to meet us with grace.

Closing Charge: Breach the Line

You have seen the enemy's most insidious pre-conflict strategy: deterrence. He knows that the most decisive battle is the one that is never fought. Like the superpowers of the Cold War who relied on the threat of annihilation to maintain a tense peace, Satan seeks to keep you in a state of spiritual paralysis. He erects a blockade around the presence of God; a seemingly impenetrable wall built from the noise of distraction and the poison of shame. His goal is to convince you that the cost of approaching your Commander is too high, that you are too busy, too flawed, or too insignificant to enter the throne room.

He wants you to be a Martha, so consumed by the good and necessary tasks of life that you miss the one thing that is truly essential: sitting at the feet of Jesus. He wants you to be the prodigal son, frozen in the pigsty, convinced that the chasm of your failure is too wide to cross. His entire deterrence campaign is designed to make you accept a ceasefire in a war you were created to win, to keep you from the very source of your power, wisdom, and strength.

But you have been given the counterintelligence. You know that this blockade is an illusion. The noise of the world is a jamming signal that can be silenced by discipline. The poison of shame is an accusation that has been nullified by the blood of Christ. God's throne is not a fortified bunker; it is a throne of grace, and your access has been permanently secured by your Savior.

Therefore, your charge is to reject the enemy's terms. You must refuse to be deterred.

Seize the Initiative: Do not wait for a convenient time to seek God. It will never come. You must, like a disciplined soldier, create it. Schedule your time with God and guard it fiercely. This is a deliberate act of war against the strategy of distraction.

Reject the Propaganda: When the accuser whispers that you are unworthy, you will counter assault with the truth of the Gospel. You will not argue your merits; you will plead the blood of Christ. You will silence the lie of shame with the truth of your unconditional acceptance.

Breach the Line with Confidence: The enemy wants you to hesitate. God commands you to approach His throne with confidence. Obeying this command is an act of spiritual defiance. Every time you open your Bible when you feel distracted, every time you fall to your knees in prayer when you feel ashamed, you are walking straight through the enemy's psychological defenses and proving his deterrents powerless.

Will you remain outside the gates, convinced by the enemy's psychological operations that you cannot enter? Or will you, like Mary, choose the one thing that is better? Will you, like Nehemiah, refuse to be distracted from your great work? Will you, like the prodigal son, begin the walk home, confident that your Father is already running to meet you?

The blockade is a bluff. The throne room is open. The Commander is waiting.

Engage.

Chapter 6

Preparation & Mobilization (Positioning His Forces)

Military Warfare:
Forging the Instrument of Victory

A STRATEGY, NO matter how brilliant, is a useless piece of paper without a force that is adequately trained, properly equipped, and effectively mobilized to execute it. This is the critical phase of readiness, where a nation transforms its raw potential – its manpower and industrial base – into a sharp, lethal, and sustainable fighting instrument. An unprepared army, no matter its size, is not a military force; it is merely a uniformed crowd marching toward disaster.[1]

The process of preparation is relentless and multi-faceted. First and foremost is **training.** Soldiers must be drilled until their skills become muscle memory. From marksmanship and first aid to complex combined-arms maneuvers, training builds the individual proficiency and unit cohesion necessary to withstand the chaos of combat. Modern militaries conduct massive, realistic exercises – like NATO's Cold War-era REFORGER exercises or the multinational RIMPAC naval drills – to test their doctrines, stress their command structures, and ensure their forces can operate seamlessly as a single, coordinated machine.[2]

Equally critical is **logistics.** The old military axiom states, "Amateurs talk strategy; professionals talk logistics." An army, it is said, marches on its stomach.[3] But it also marches on its fuel, its ammunition, its spare parts, and its medical supplies. The "tail" of logistical support is what enables the "tooth" of the combat arms to fight. Mobilization involves not just the movement of troops, but the creation of vast and complex supply chains that can sustain a force thousands of miles from home. A failure in logistics can halt the most powerful offensive in its tracks.

Finally, there is the **positioning of forces.** This is the physical act of mobilizing troops and equipment into the theater of operations, setting the conditions for the battle to come. It is a monumental undertaking, a clear sign of intent, and the final step before the initiation of hostilities.

Case Study: The German Rearmament (1930s)

A chilling example of long-term, national-level preparation can be seen in Germany's rearmament throughout the 1930s.[4] Following its defeat in World War I and the strict limitations imposed by the Treaty of Versailles, Germany was a militarily neutered state. However, upon rising to power, Adolf Hitler initiated a massive and clandestine program of preparation and mobilization designed to build the most powerful army in the world.

This was not just about building tanks and planes. It was a total societal mobilization for war. The Hitler Youth indoctrinated and physically conditioned an entire generation of boys for future military service. The German economy was reoriented toward military production.

German engineers, forbidden from designing tanks, secretly designed "agricultural tractors" that bore a suspicious resemblance to armored vehicles.[4] Pilots were trained in civilian flying clubs. This decade-long, deliberate preparation forged the Wehrmacht, the force that would later unleash the Blitzkrieg on Europe. By the time Germany invaded Poland in 1939, its forces were not only well-equipped but also highly trained and ideologically prepared for the conflict they had long been planning.

Case Study: The Buildup to Operation Overlord (1944)

The Allied preparation and mobilization for the D-Day invasion was arguably the greatest logistical and organizational feat in military history.[5] The decision to open a second front in Europe was made years before the actual landing, and the intervening time was dedicated to a staggering buildup of forces in the United Kingdom.

By the spring of 1944, southern England had been transformed into a massive military camp. Over 1.5 million American soldiers were shipped across the Atlantic, joining British, Canadian, and other Allied forces.[6] The sheer volume of materiel was breathtaking: thousands of tanks, tens of thousands of vehicles, and millions of tons of supplies, from ammunition and fuel to food and medical equipment, had to be produced, transported, and staged.

The training was intense and highly specific. Units rehearsed amphibious landings on British beaches that mimicked the terrain of Normandy. Paratroopers conducted countless practice jumps. Engineers trained on how to clear obstacles under fire. This massive mobilization was not just about accumulating force; it was about positioning that force for a single, decisive moment. When the order to launch the invasion was finally given, it set in motion a perfectly prepared and precisely positioned force, the culmination of years of planning and preparation. The success of the Normandy landings was secured long before the first soldier stepped onto the beach; it was secured in the factories of Detroit, the training grounds of England, and the minds of the logisticians who made it all possible.[7]

These examples reveal a fundamental law of warfare: victory is the direct result of superior preparation. An army that has trained harder, supplied itself better, and positioned itself more effectively has already set the conditions for its success.

Spiritual Warfare: Preparing the Battlefield

Just as a nation mobilizes its industry and trains its troops long before an invasion, our enemy is engaged in a constant, long-term campaign of preparation and mobilization. He is not merely waiting for an opportunity to attack; he is actively shaping the environment to make his future assaults more effective. He is positioning his forces, strengthening temptation, and preparing the cultural battlefield to ensure that when he does strike, the conditions are overwhelmingly in his favor.

This spiritual mobilization is a subtle but powerful form of warfare. The enemy understands that a direct assault on a well-prepared believer is difficult. It is far easier to first degrade the entire spiritual environment, normalizing sin and weakening the collective resolve of God's people. He works to transform the cultural landscape from hostile territory into a welcoming staging ground for his operations. As the Apostle Paul states, before coming to Christ, we "followed the ways of this world and of the ruler of the kingdom of the air, the spirit who is now at work in those who are disobedient" (Ephesians 2:2). Satan prepares the battlefield by ensuring the "ways of this world" are as seductive and spiritually debilitating as possible.

His methods of preparation are threefold:

Normalizing Sin: He works to desensitize an entire culture to sin, taking behaviors that were once considered shameful and rebranding them as normal, enlightened, or even virtuous. Through media, entertainment, and academia, he relentlessly promotes a narrative that glorifies "the lust of the flesh, the lust of the eyes, and the pride of life" (1 John 2:16). What was once whispered in darkness is now shouted from the rooftops and celebrated in parades. This cultural shift creates an atmosphere where constant exposure to sin erodes our spiritual defenses and makes temptation seem less dangerous.

Creating and Distributing "Weapons": The enemy actively inspires the creation of addictive substances, behaviors, and technologies that serve as weapons against the soul. From the development of highly potent narcotics to the creation of infinitely scrolling, dopamine-hijacking social media apps, he promotes tools designed to enslave. He ensures that sinful content, particularly pornography, is not only readily accessible but aggressively pushed to every corner of the internet. This is the logistical arm of his mobilization: ensuring the "ammunition" for temptation is stockpiled and easily available to all.

Fueling the Flesh: Ultimately, his preparation aims to fuel our own sinful desires. James 1:14 states that we are tempted when we are "dragged away by [our] own evil desire and enticed." Satan's mobilization is designed to make that "evil desire" as strong and as well-fed as possible. He creates environments that cater to our specific weaknesses, providing endless opportunities to indulge in the very things that draw us away from God.

Case Study: The Seduction of Israel at Peor (Numbers 25 & 31)

One of the most chilling examples of the enemy strategically preparing the battlefield is the advice given by the prophet Balaam. After failing to curse Israel directly, Balaam provided the Moabite king Balak with a devastating new strategy. He couldn't defeat Israel with spiritual power, so he would advise the Moabites on how to make Israel defeat themselves.

The Bible later reveals Balaam's counsel: he taught Balak to entice the Israelite men with Moabite women, luring them into sexual immorality and, critically, into the worship of their idol, Baal of Peor (Numbers 31:16). This was not a random encounter. It was a deliberate, prepared temptation. The Moabites mobilized their women and positioned them as a spiritual weapon. They prepared a feast and an idolatrous ritual. They created an environment specifically designed to exploit the weaknesses of the Israelite men.

The strategy was horrifically successful. The Israelite men fell into the trap, and a plague from the Lord killed 24,000 of them. Israel's army was decimated not by enemy swords, but by a pre-planned, mobilized temptation that they were not prepared to resist. The enemy had successfully prepared the battlefield to turn their own desires against them.

Counterintelligence:
Personal Readiness and Mobilization

If the enemy is in a constant state of preparation and mobilization, then the believer cannot afford to live in a state of spiritual peacetime. We must engage in our own continuous process of preparation, training, and equipping. Our counterintelligence is not a passive waiting game, but the active discipline of forging our souls into instruments of victory, ready for the conflict we know is coming. This is the essence of Paul's command to "put on the full armor of God" (Ephesians 6:11) – an act of deliberate, daily readiness.

Our personal mobilization can be understood through the same lens as military preparation:

1. Personal Training (Building Spiritual Disciplines)
A soldier's effectiveness is determined by his training. Our spiritual training is accomplished through the consistent practice of spiritual disciplines. These are not religious chores; they are the drills and exercises that build our spiritual strength, endurance, and proficiency.

Prayer: This is our daily communication with High Command. It is where we receive our orders, report on our status, and call for fire support. A disciplined prayer life keeps us connected and spiritually alert.

Study of the Word: This is our intelligence briefing and our training manual. Immersing ourselves in Scripture equips us to recognize the enemy's tactics, understand our Commander's strategy, and wield the "sword of the Spirit."

Fasting: This is an advanced training exercise. Fasting disciplines the flesh, sharpens our spiritual discernment, and increases our dependence on God. It is a deliberate act of choosing the spiritual over the physical, preparing us for times of intense trial.

2. Spiritual Logistics (Managing Your Environment)

A professional army pays obsessive attention to its supply lines and its environment. Our spiritual logistics involve being ruthlessly aware of the influences we allow into our lives. We must actively manage our environment to protect ourselves from the enemy's mobilized temptations.

Media and Information Diet: You must be the commanding officer of your own mind. This means being highly selective about the media you consume – the shows, the music, the websites, the social media feeds. If what you are consuming is normalizing sin, fueling discontent, or promoting godlessness, you are allowing the enemy to run a supply line of propaganda directly into your camp.

Cultivating Godly Relationships: Your closest relationships are a key part of your spiritual environment. Surround yourself with fellow believers who will encourage your faith, hold you accountable, and fight alongside you. Actively distance yourself from relationships that consistently pull you toward compromise and sin.

Sanctifying Your Space: Actively seek out and create environments that promote spiritual growth. Make your home a place of peace and worship. Spend time in nature to connect with the Creator. Be an active part of a vibrant, Bible-teaching church. These are your fortified positions.

3. Mobilization for Action (Living a Life of Purpose)

Preparation is not an end in itself. We train and equip ourselves so that we can be mobilized for the mission. This means living a life of active, obedient purpose. God has not called us to sit in a fortified bunker, but to advance His Kingdom. Every act of service, every time you share your faith, every act of love and generosity is a deployment. It is an offensive maneuver that pushes back the darkness. A believer who is actively engaged in the mission of God – using their gifts, serving others, and sharing the Gospel – is a mobilized soldier, and a far more difficult target for the enemy to attack than one who is sitting idly in the barracks.

Biblical Assault:
After-Action Reports on Readiness

The distinction between victory and defeat in Scripture often comes down to one critical factor: preparation. Those who were prepared for the moment of crisis stood firm, while those who were unprepared were swept away.

Case Study: The Parable of the Ten Virgins (Matthew 25)

Jesus Himself gives us the ultimate parable on the importance of preparation and readiness. The story is simple: ten virgins are waiting for the bridegroom. All ten have lamps, but only five are wise enough to bring extra oil.

The State of Readiness: The lamps represent their outward profession of faith, but the oil represents their inner, personal preparation. The five wise virgins understood that readiness is not a one-time event, but a state of continuous preparation. They anticipated a potential delay and mobilized the necessary resources (the extra oil) beforehand. The five foolish virgins were unprepared. They had the appearance of readiness (their lamps), but they lacked the substance (the oil). They failed to prepare for anything beyond the immediate future.

The Moment of Crisis: The bridegroom was delayed, and the call came at midnight. This is the moment of crisis, the unexpected trial. All ten lamps began to go out. The foolish virgins, in their panic, realized their lack of preparation. Their plea to the wise – "Give us some of your oil" – reveals a fatal misunderstanding. Personal readiness cannot be transferred at the last minute.

The Strategic Consequence: While the foolish virgins were away trying to buy oil, the bridegroom arrived, and the door was shut. Their lack of preparation meant they were absent at the most critical moment. The chilling final words of the parable serve as a universal military and spiritual principle: "Therefore keep watch, because you do not know the day or the hour" (Matthew 25:13). Preparation is not optional; it is the prerequisite for entry.

Case Study: David vs. Goliath (1 Samuel 17)

The iconic clash between David and Goliath is often seen as a story of miraculous, spontaneous deliverance. In reality, it is a story of a prepared warrior confronting an unprepared army.

The Unprepared Force: For forty days, the army of Israel, including King Saul, was completely paralyzed by fear. They were a mobilized army, equipped with swords and spears, but they were psychologically and spiritually unprepared for this specific type of threat. Their fear made their armor and weapons useless.

The Prepared Warrior: David, on the other hand, had been in training for this moment his entire life, though he did not know it.

He was Technically Prepared: He had spent countless hours in the fields protecting his sheep, mastering the sling – a simple but deadly weapon. When he picked up five smooth stones, he was not grabbing random rocks; he was selecting the ammunition he had trained with for years.

He was Experientially Prepared: He recounted to Saul how he had fought and killed both a lion and a bear to save his flock. He had faced terrifying enemies before and had seen God deliver him. He had a history of successful operations that fueled his confidence.

He was Spiritually Prepared: Most importantly, David's faith was not theoretical. It was battle-tested. His confidence was not in himself, but in his Commander. His famous battle cry was a declaration of his spiritual readiness: "You come against me with sword and spear and javelin, but I come against you in the name of the LORD Almighty, the God of the armies of Israel, whom you have defied" (1 Samuel 17:45).

David won because he was the only one on the battlefield who was truly prepared for the fight. His victory was the culmination of years of quiet, faithful preparation, proving that readiness is the key that unlocks divine power.

Closing Charge: The Call to Readiness

You have seen the enemy's doctrine of preparation. He is not idle. He is actively mobilizing his forces, shaping the culture, and preparing the battlefield to exploit your weaknesses. He is positioning temptation, normalizing sin, and working to ensure that the environment you live in is perfectly suited for his offensive operations. He is counting on you to be distracted, undisciplined, and unprepared.

But the call of your Commander is not to a life of spiritual passivity. It is a call to a state of constant readiness. You are to be a wise virgin with oil in your lamp. You are to be a David who has trained in the fields before you face the giant. Your spiritual disciplines – prayer, study, fellowship, fasting – are not religious rituals; they are your training exercises. Your management of your environment – what you watch, who you listen to, where you go – is your spiritual logistics. Your active service in the kingdom is your mobilization.

Do not be deceived into believing you can cram for the final exam. Do not think you can afford to be spiritually lazy today and hope to be strong in the battle tomorrow. The crisis will come at midnight, when you least expect it. The giant will not wait for you to feel ready.

Therefore, embrace the discipline of preparation. Take your training seriously. Guard your supply lines. Know your weapon, the Sword of the Spirit. Live in a state of readiness, not out of fear, but out of a deep and abiding confidence that the one who is prepared is the one God will use. The battle is coming. The question is not will you be called, but will you be ready?

Prepare for **war**.

Phase II

The Initial Phase
(The Assault)

THE RECONNAISSANCE IS complete. The enemy's preparations are finalized. His objectives have been defined, and his forces are mobilized. The time for silent preparation is over. The hour for the initial assault has arrived.[1]

This is the phase of the war where the enemy seizes the initiative. His goal is no longer just to study, but to strike – to test our defenses, exploit our weaknesses, and gain a decisive early advantage. The attacks that follow are not random; they are a coordinated campaign designed to achieve strategic shock and awe in our spiritual lives.[2]

In the chapters that follow, we will analyze the enemy's opening moves. He will seek to achieve immediate air superiority by controlling the narrative that shapes our thoughts and culture.[3] He will attempt to suppress our defenses by silencing the truth of the Gospel. He will launch psychological operations designed to demoralize and disorient us, and he will execute rapid, targeted strikes against our most vulnerable points.[4]

The preliminary maneuvers are over.

The ***assault*** has begun.

Chapter 7

Gaining Air Superiority (Controlling the Narrative)

Military Warfare: Dominating the Skies

IN THE DOCTRINE of modern warfare, the sky is the ultimate high ground. The ability to control the airspace over a battlefield – to achieve air superiority – is often the single most critical factor determining victory or defeat.[1] An army that commands the skies can operate with impunity, striking enemy targets at will, gathering intelligence without interference, and resupplying its own forces with freedom. Conversely, an army that has lost control of the air is blind, vulnerable, and constantly under threat. Every movement is watched, every supply convoy is a target, and every defensive position is at risk of annihilation from above. Gaining air superiority is the essential first move of the initial assault, the act of kicking in the door to enable all subsequent ground operations.

Achieving this dominance is a violent and systematic process. It involves the targeted destruction of the enemy's air force and its supporting infrastructure. The campaign begins with strikes against enemy airfields, destroying aircraft on the ground before they can ever take off. It targets radar installations and anti-aircraft missile sites, blinding the enemy and clearing safe corridors for friendly aircraft. Finally, it involves aggressive air-to-air combat, with fighter jets sweeping the skies to hunt down and eliminate any remaining enemy planes. The goal is not just to win a dogfight; it is to totally dismantle the enemy's ability to contest the airspace. Once air superiority is achieved, the war fundamentally changes. The commander who controls the sky controls the battlefield.

Case Study: The Six-Day War
(1967) – The Preemptive Strike

No conflict in modern history demonstrates the decisive power of gaining immediate air superiority more starkly than the Six-Day War. Facing the combined and numerically superior armies of Egypt, Syria, and Jordan, Israel knew that a prolonged, multi-front war would be unsustainable. Its survival depended on a swift, paralyzing blow delivered in the opening hours of the conflict.

On the morning of June 5, 1967, the Israeli Air Force (IAF) launched Operation Focus.[2] In a series of massive, meticulously planned surprise attacks, waves of Israeli jets flew low to avoid radar and struck Egyptian airfields across the Sinai and Egypt proper. In less than three hours, the IAF had annihilated the vast majority of the Egyptian Air Force, destroying hundreds of combat aircraft while they were still parked on the tarmac. The IAF then turned its attention to the air forces of Jordan, Syria, and Iraq, achieving similar devastating results.

By the end of the first day, Israel had achieved complete and total air superiority over the entire region.[3] This opening move was strategically decisive. With the skies clear of threats, the IAF became, in effect, flying artillery for the Israeli ground forces. Israeli tanks and infantry advanced across the Sinai and into the Golan Heights with devastating air support, while the Egyptian and Syrian armies, bereft of any air cover, were exposed and relentlessly pounded from above. The war was effectively won in the first few hours, not by a ground battle, but by a brilliant and ruthless campaign to seize control of the sky.

Case Study: The Battle of Britain
(1940) – A Defensive Victory

While the Six-Day War shows the power of an offensive campaign for air superiority, the Battle of Britain demonstrates the critical importance of *denying* it to the enemy. After the fall of France in 1940, Great Britain stood alone against the might of Hitler's Germany. A German land invasion of Britain, codenamed Operation Sea Lion, was imminent. But the German High Command knew such an invasion could not succeed without first destroying Britain's Royal Air Force (RAF).

The Luftwaffe, the German air force, launched a massive air campaign to shatter the RAF and clear the skies over the English Channel.[4] For months, hundreds of German bombers and fighters clashed with the outnumbered RAF in the skies over southern England. The British, however, had a decisive technological and strategic advantage. Their newly developed radar system gave them early warning of incoming German raids, allowing them to scramble their fighters to intercept them efficiently.[5] The command and control system, known as the Dowding System, allowed RAF commanders to manage the battle with a level of coordination the Germans could not match.

The pilots of RAF Fighter Command, flying their iconic Spitfires and Hurricanes, fought with incredible courage and tenacity. Though stretched to their limit, they never broke. They inflicted unsustainable losses on the Luftwaffe, making the cost of the air campaign too high for Germany to bear. By failing to achieve air superiority, the Germans could not provide the necessary air cover for an invasion fleet.[6] Operation Sea Lion was postponed indefinitely and ultimately cancelled. The RAF's defensive victory in the Battle of Britain did not win the war, but it ensured that Britain could not lose it. By successfully denying air superiority to the enemy, they saved their nation from invasion and preserved it as the crucial unsinkable aircraft carrier from which the Allies would eventually launch their return to Europe.

Spiritual Warfare:
Controlling the Narrative

In the spiritual war, the sky is not made of air and clouds; it is the intellectual and cultural atmosphere we breathe every day. It is the realm of ideas, beliefs, narratives, and worldviews. Just as a military force seeks to dominate the physical airspace, our enemy's first move in any major assault is to achieve **narrative superiority.** His goal is to control the stories our culture tells, the ideas it celebrates, and the truths it accepts. He knows that if he can control the "airwaves" of a society, he can shape the thoughts, beliefs, and desires of its people long before they ever encounter the truth of the Gospel.

This is a strategic, preemptive campaign. Satan seeks to create a cultural environment that is fundamentally hostile to the message of Christ. He wants to ensure that by the time a believer attempts to share the Gospel, the listener's mind has already been fortified with arguments against it. He wants the core tenets of the Christian faith – sin, repentance, judgment, salvation through a crucified Savior – to seem not just untrue, but foolish, archaic, and offensive. By controlling the narrative, he aims to win the war in the mind before the battle for the soul can even be waged.

His tactics for achieving this narrative dominance are sophisticated and pervasive:

Propaganda and Misinformation: The enemy relentlessly disseminates false ideologies through influential channels: academia, media, and entertainment. He promotes worldviews like secular humanism, moral relativism, and naturalism, which are designed to exclude God from the picture entirely. He twists history, redefines morality, and spreads misinformation that paints the Church as a force for oppression rather than liberation. As the Apostle Paul warned, the enemy uses "all sorts of displays of power through signs and wonders that serve the lie" (2

Thessalonians 2:9), creating a convincing but utterly false picture of reality.

Controlling the Language: A key part of controlling the narrative is controlling the language. The enemy works to redefine words to suit his agenda. "Truth" becomes subjective ("your truth"). "Love" is redefined as unqualified affirmation of any behavior. "Tolerance" is twisted to mean acceptance of all ideas except for exclusive biblical truth. "Justice" is divorced from righteousness and attached to worldly ideologies. By changing the meaning of words, he changes the nature of the conversation itself.

Influencing Cultural Trends: He seeks to shape the stories our culture tells through movies, music, and art. He normalizes and glorifies what God has forbidden, making sin seem glamorous, exciting, and liberating. He creates a cultural current that constantly pulls people away from biblical morality, making a godly life seem like an act of swimming upstream against an overwhelming tide.

Leveraging Social Media: In the modern era, social media is his prime instrument for achieving narrative superiority. It allows for the rapid spread of viral misinformation. Its algorithms create powerful echo chambers, reinforcing false beliefs and isolating users from opposing viewpoints. It fosters a culture of outrage, envy, and division, making reasoned, grace-filled dialogue nearly impossible.

Case Study: Paul at the Areopagus (Acts 17)

The Apostle Paul's visit to Athens provides a powerful case study of a direct engagement in the battle for narrative superiority. Athens was the intellectual capital of the world, a city whose "airspace" was thick with competing philosophies and ideologies. The Athenians, the Bible notes, "spent their time doing nothing but talking about and listening to the latest ideas" (Acts 17:21). It was a battlefield of narratives.

When Paul began to preach, he was brought to a meeting of the Areopagus, the city's philosophical high council. Here, he stood face-to-face with the dominant narratives of his day, represented by the Epicurean and Stoic philosophers. These were not just intellectual positions; they were comprehensive worldviews that sought to explain the nature of reality and the purpose of human life.

Paul's response was a perfect display of spiritual counter-narrative warfare.

He Identified Their Existing Narrative: He didn't begin with a frontal assault. He started by showing he understood their "airspace." "People of Athens! I see that in every way you are very religious. For as I walked around and looked carefully at your objects of worship, I even found an altar with this inscription: TO AN UNKNOWN GOD." He found a crack in their existing narrative – an admission of spiritual ignorance.

He Commandeered Their Narrative: He then brilliantly commandeered their own altar and used it as a launchpad for the Gospel. "So you are ignorant of the very thing you worship – and this is what I am going to proclaim to you."

He Deconstructed Their False Narratives: He systematically dismantled their flawed worldviews. Against their pantheon of distant, man-made gods, he presented the one true Creator who made the world and everything in it, who is not far from any one of us.

He Deployed the True Narrative: Finally, having cleared the airspace of their false assumptions, he deployed the full truth of the Gospel: God has commanded all people everywhere to repent, and He has given proof of this by raising a man from the dead.

In this single encounter, Paul engaged in a direct battle for control of the narrative, demonstrating how to counter the prevailing lies of the culture with the unshakeable truth of God.

He proved that even in the most hostile intellectual airspace, the Gospel, when proclaimed with wisdom and boldness, can penetrate the enemy's defenses.

Counterintelligence: Seizing the Narrative High Ground

In modern information warfare, a nation cannot defend against enemy propaganda simply by ignoring it. To counter a hostile narrative, one must engage in a sophisticated and proactive counterintelligence campaign. This involves not only identifying and exposing the enemy's disinformation but also dominating the information space with a more compelling, truthful, and powerful narrative of one's own. It requires training citizens and soldiers to become critical thinkers, to question the sources of information, and to recognize the subtle markers of psychological manipulation. The goal is not just to defend against lies, but to make the truth so clear and resonant that the lies lose their power to persuade.

If the enemy's primary assault is to achieve narrative superiority by controlling the cultural airwaves, then our counterintelligence must be a deliberate campaign to reclaim that airspace. We cannot be passive consumers of the culture around us; we must become active, discerning analysts and bold broadcasters of the truth. We are called to demolish the enemy's arguments and ideological strongholds by taking every thought captive and making it obedient to the reality of Christ (2 Corinthians 10:5).

Directive 1: Establish Your Invariant Truth Standard (The Command Doctrine)

Every military operates on a set of core doctrines – the fundamental principles that guide all actions. This doctrine is the ultimate standard against which every order and every piece of intelligence is measured. For the believer, our unchanging, non-negotiable command doctrine is the Word of God. To fight a war of narratives without being deeply and personally grounded in

Scripture is the equivalent of trying to navigate a battle without a map, a compass, or a radio link to headquarters.

This requires more than a casual acquaintance with the Bible. It demands a commitment to knowing the "whole counsel of God" (Acts 20:27). We must become so saturated in biblical truth that we can instantly recognize a counterfeit.

Know God's Character: The enemy's narratives often subtly misrepresent God's character, painting Him as either a cosmic tyrant or an indulgent grandfather. A deep knowledge of Scripture reveals His true nature: perfectly just, infinitely merciful, radically holy, and unconditionally loving.

Understand Your Identity: The culture will try to define you by your feelings, your desires, your career, or your group affiliation. Scripture provides your true identity: you are a new creation, a child of God, an ambassador for Christ, and a temple of the Holy Spirit.

Grasp the True Gospel: The enemy will always try to promote a "different gospel" (Galatians 1:6) – one of works, of self-improvement, or of universal affirmation. You must be unshakably rooted in the truth of salvation by grace alone, through faith alone, in Christ alone. This biblical doctrine is the filter through which all cultural narratives must pass.

Directive 2: Practice Critical Consumption (Active Signal Interception)

An intelligence analyst doesn't just passively listen to enemy radio traffic; they actively intercept, decode, and analyze it. They question its source, its motive, and its intended effect. We must apply this same critical discipline to every piece of information the world sends our way. As Proverbs 14:15 warns, "The simple believe anything, but the prudent give thought to their steps."

This means engaging your mind, not turning it off, when you interact with culture. As you watch a movie, read an article, or scroll through social media, become an active analyst. Ask the counterintelligence questions:

What is the underlying worldview being promoted here? Is it that humanity is basically good? That truth is relative? That fulfillment is found in material possessions or romantic love?

What is this narrative trying to make me feel? Fear? Envy? Anger? Lust? A sense of victimhood?

What is being glorified, and what is being mocked? Is biblical morality portrayed as foolish and oppressive? Is sin portrayed as liberating and sophisticated?

Who benefits from me believing this message? By practicing this level of discernment, you move from being a passive target of enemy propaganda to an active agent who can identify and neutralize narrative attacks in real-time.

Directive 3: Launch a Counter-Offensive (Broadcasting on God's Frequency)

It is not enough to simply reject lies in the privacy of our own minds. To reclaim the airspace, we must actively broadcast the truth. We are called to be salt and light – preservative and illuminating agents in a decaying, dark culture. Your life is a broadcast tower, and you have a choice about which signal you will transmit.

Your Testimony: Sharing what God has done in your life is a powerful counter-narrative to the world's message of hopelessness. It is a real-world report of the truth and power of the Gospel.

Your Work: Excellence, integrity, and a servant's heart in your secular workplace broadcasts a powerful message about the character of your King.

Your Family: Raising your children to know and love the Lord, to think critically, and to understand their identity in Christ is one of the most powerful long-term strategies for taking back cultural ground.

Your Words: In your daily conversations and your online interactions, you can choose to counter the prevailing narratives of outrage and division with grace, wisdom, and truth.

Every act of bold, grace-filled truth-telling is a direct assault on the enemy's control of the narrative. It is like setting up a pirate radio station that broadcasts the frequency of the Kingdom of God in the middle of enemy-occupied territory.

Directive 4: Secure Your Local Airspace (Home and Church Fortification)

While the narrative war rages on a global scale, it is fought most intensely in our local environments. You must take responsibility for the "airspace" over your own home and your own church.

Curate Your Media Diet: As the gatekeeper of your home, especially if you have children, you have the authority and responsibility to filter the propaganda coming through your screens. Make intentional choices. Have conversations about the worldviews being promoted in the shows you watch.

Ensure Doctrinal Purity in Your Church: Support and pray for your pastors and elders. Ensure your church is a place where the Word of God is taught without apology and where cultural narratives are examined through the lens of Scripture, not the other way around. A biblically sound church is a vital command-and-control center in the information war.

The battle for the sky – the battle for the narrative – is the first and often most decisive stage of the assault. By becoming a people deeply rooted in God's Word, critically discerning of the world's messages, and bold in proclaiming the truth, we refuse to cede the high ground. We plant the flag of our King in the cultural airspace and declare that the truth of the Gospel will not be silenced.

Biblical Assault: After-Action Reports on Narrative Warfare

The battle for the narrative is as old as faith itself. Throughout Scripture, we see God's prophets, apostles, and even His own Son engaging in direct confrontations with the prevailing lies of their culture. These are not simple debates; they are spiritual assaults designed to demolish false worldviews and seize the narrative high ground for the Kingdom of God. By studying these campaigns, we learn how to counter the enemy's propaganda with the overwhelming power of divine truth.

Case Study: Elijah on Mount Carmel (1 Kings 18) – A Decisive Air Strike

During the reign of King Ahab and his wicked wife Jezebel, the dominant narrative in Israel was that Baal, the pagan storm god, was in control. The worship of the one true God had been suppressed, His prophets were hunted, and the "airspace" over the nation was thick with idolatry. The enemy had achieved near-total narrative superiority. Elijah, operating as a lone agent, launched a daring assault to reclaim it.

The Enemy's Narrative: "Baal is the god who brings the rain and controls the fortunes of this land. Yahweh is either powerless or irrelevant."

The Biblical Assault: Elijah did not engage in a subtle war of ideas. He staged a decisive, public, winner-take-all showdown on Mount Carmel. He proposed a test that was simple and undeniable: two altars, two bulls, and no fire. The god who answered with fire from heaven would be declared the true God. This was a direct challenge to Baal on his own turf – the god of lightning and fire. After the prophets of Baal spent hours chanting, crying, and gashing themselves with no result, Elijah stepped forward. He repaired the Lord's altar, drenched it in water to make the miracle undeniable, and prayed a simple, powerful prayer.

The Result: "Then the fire of the LORD fell and burned up the sacrifice, the wood, the stones and the soil, and also licked up the water in the trench" (1 Kings 18:38). The display was so overwhelming, so absolute, that it shattered the enemy's narrative in an instant. The people fell prostrate and cried out, "The LORD – he is God! The LORD – he is God!" Elijah seized narrative control not by arguing, but by demonstrating the superior power of his Commander.

Case Study: Jesus vs. The Pharisees (Matthew 5) – A Sustained Campaign

The Pharisees of Jesus' day controlled the dominant religious narrative in Israel. They taught that righteousness was a matter of meticulous external observance of the Law and their man-made traditions. Theirs was a narrative of pride, exclusion, and self-righteousness. Jesus's entire public ministry was a sustained assault on this false narrative.

The Enemy's Narrative: "Righteousness is achieved by what you do externally. If you follow the rules, you are right with God."

The Biblical Assault: In the Sermon on the Mount, Jesus launched a devastating series of narrative strikes. He did not abolish the Law; He exposed the Pharisees' shallow interpretation of it. With each repetition of the phrase, "You have heard that it was said...," He identified their flawed narrative. With each "But I tell you...," He replaced it with the deeper, true narrative of the Kingdom.

They said, "Do not murder." Jesus assaulted this by revealing the root issue: "anyone who is angry with a brother or sister will be subject to judgment."

They said, "Do not commit adultery." Jesus assaulted this by revealing the root issue: "anyone who looks at a woman lustfully has already committed adultery with her in his heart." He systematically dismantled their external, performance-based system and replaced it with the Kingdom narrative of internal heart transformation. This was a revolutionary assault that redefined the very concept of righteousness.

Case Study: Shadrach, Meshach, and Abednego (Daniel 3) – Narrative Defiance

King Nebuchadnezzar, the most powerful man on earth, erected a giant golden idol and commanded all his subjects to bow down and worship it. This was an exercise in absolute narrative control.

The Enemy's Narrative: "The power of the state is absolute, and loyalty is demonstrated through unified, mandatory worship of its chosen idol." To defy this was to defy the king and face certain death.

The Biblical Assault: The assault from the three young Hebrew men was not one of violence, but of defiant stillness. Their response to the king is one of the most powerful counter-narratives ever spoken: "King Nebuchadnezzar, we do not need to defend ourselves before you in this matter... we want you to know, Your Majesty, that we will not serve your gods or worship the image of gold you have set up" (Daniel 3:16-18). This was a declaration that there was a higher authority than the king and a truer narrative than the one he was imposing.

The Result: After their miraculous survival in the fiery furnace, the most powerful man in the world was forced to abandon his own narrative and adopt theirs. Nebuchadnezzar issued a new decree, declaring that "no other god can save in this way" (Daniel 3:29). The bold defiance of three men successfully forced the entire empire to acknowledge the superior power and reality of their God. They proved that a false narrative, even one backed by the threat of death, cannot stand against the truth.

Closing Charge: Seize the High Ground

You have seen the enemy's opening move in his assault phase: the battle for air superiority. He knows that the war for your soul is first fought in the airspace of your mind and your culture. His objective is to achieve narrative dominance, to fill the airwaves with a fog of lies, half-truths, and distractions until the truth of the Gospel seems distant and irrelevant. He wants to control the story, because he who controls the story controls the belief. He seeks to deconstruct reality, redefine truth, and create a cultural atmosphere so hostile to God that faith cannot survive.

He will use the media to glorify sin and mock righteousness. He will use academia to teach that God is a myth and that humanity is the master of its own destiny. He will use the echo chambers of social media to amplify division and outrage. This is his air campaign, a relentless bombardment designed to soften the target, demoralize the populace, and ensure that his ground invasion of your heart meets with little resistance.

But you have not been left defenseless against this onslaught. You are not a civilian, cowering in a shelter. You are a warrior, and you have been commanded to engage in this fight. Your Commander has not ceded the high ground. The truth is the most powerful narrative in the universe, and you have been entrusted with it.

Therefore, your charge is to counter his air campaign with one of your own.

Know Your Doctrine: You cannot fight a war of ideas if you are ignorant. You must be so deeply rooted in the truth of Scripture that you can instantly identify the enemy's propaganda. Make the Word of God your non-negotiable reality.

Think Critically: Do not be a passive consumer of the culture. Be an active analyst. Question the narratives you are being fed. Identify the underlying worldviews. Discern the emotional manipulation at play. A discerning mind is a hardened target.

Broadcast the Truth: Your life is a signal. Counter the enemy's broadcast of lies with your broadcast of God's truth. Share your testimony. Live with integrity. Speak with grace. Love your neighbor. Every act of authentic faith is a direct assault on the enemy's narrative control.

Defend Your Local Airspace: Take responsibility for the narrative being broadcast in your own home. Curate what your family consumes. Teach your children to think biblically and critically. Fortify your local church as a bastion of truth where the "whole counsel of God" is proclaimed without fear or compromise.

Will you allow the enemy to dominate the skies of your mind? Will you let his propaganda shape your view of the world, of God, and of yourself? Or will you, like Elijah, call down the fire of truth to expose the impotent lies of the age? Will you, like Paul in Athens, boldly proclaim the true narrative in the heart of the enemy's intellectual territory?

The battle for the narrative is raging. It is time to clear the skies. It is time to seize the high ground.

Shane Cunningham

Chapter 8

Information Warfare (Propaganda & Psychological Operations)

Military Warfare: The Battle for the Mind

IN THE MODERN military lexicon, the battlefield is understood to be multidimensional.[1] It exists not only on land, at sea, and in the air, but also in the minds of the combatants and the populations that support them. This is the domain of Information Warfare, a sophisticated and often decisive form of conflict where the primary targets are not bodies and buildings, but perceptions, beliefs, and the will to fight. It operates on the principle that an enemy who has been psychologically broken, whose morale has collapsed, and whose trust in his leaders has been shattered is an enemy already defeated. This warfare is waged with propaganda, deception, and psychological operations (PSYOPs), designed to disrupt command and control, demoralize enemy forces, and sway public opinion.[2]

Information Warfare is a broad discipline that encompasses everything from cyber-attacks on an enemy's command network to the strategic use of social media. At its core, however, are two fundamental weapons: propaganda and PSYOPs. Propaganda is the strategic dissemination of information – whether true, false, or a mixture of both – to influence an audience and advance a specific agenda.[3] It can be used to demonize the enemy, justify one's own cause, and maintain morale on the home front. Disinformation, a more sinister subset, involves the deliberate creation and spread of false information with the intent to deceive and mislead.

PSYOPs are the tactical application of these tools. They are planned operations to convey selected information and indicators to audiences to influence their emotions, motives, objective reasoning, and ultimately the behavior of governments, organizations, groups, and individuals. The ultimate goal is to win the war by capturing the enemy's mind, making physical conflict shorter, less costly, or altogether unnecessary.

Case Study: "Axis Sally" and "Tokyo Rose" in World War II

During the Second World War, the Axis powers deployed a powerful PSYOPs weapon against Allied soldiers: propaganda radio broadcasts.[4] These broadcasts were a cunning blend of entertainment and psychological manipulation, designed to attack the morale of the individual soldier at his most vulnerable point. From Germany, an American-born woman named Mildred Gillars became known to GIs as "Axis Sally." From Japan, Iva Toguri D'Aquino and other women were collectively dubbed "Tokyo Rose."[5]

Their method was strategically brilliant in its simplicity. They would play the latest popular American music, songs that were specifically chosen to evoke powerful feelings of nostalgia and homesickness. These broadcasts reminded soldiers of home, of their wives and sweethearts, of a life of peace and comfort that was a world away from the mud and terror of the front lines. Then, between the songs, their seductive and friendly voices would deliver a stream of demoralizing propaganda. They would read the names of captured or killed Allied soldiers, describe the horrors of combat in graphic detail, and relentlessly push narratives designed to create psychological distress. They suggested that their leaders were corrupt warmongers, that the war was a pointless meat grinder, and, most cruelly, that their loved ones back home were being unfaithful.

The strategic objective was clear: to sever the soldier's emotional and psychological ties to his cause and his home.[6] The goal was to erode his will to fight from the inside out by creating a corrosive cocktail of homesickness, fear, jealousy, and suspicion. It was a direct assault on morale, which military commanders know is a key component of any soldier's combat effectiveness. While the overall strategic impact of these broadcasts is debated by historians, they remain a textbook example of tactical PSYOPs aimed at the individual warrior.

Case Study: The Cold War and Radio Free Europe

While Axis Sally aimed for tactical demoralization, the use of radio during the Cold War represented a grand strategic information campaign that lasted for decades. The Soviet Union maintained its control over its satellite states in the Eastern Bloc not just with tanks and secret police, but with an "Iron Curtain" of information. They created a sealed-off society where the state-controlled media was the only source of "truth," and all outside information was banned.

In response, the United States funded and operated Radio Free Europe/Radio Liberty (RFE/RL).[7] Broadcasting from Western Europe, RFE/RL's mission was to pierce this informational Iron Curtain. It did not, for the most part, broadcast overt, heavy-handed propaganda. Instead, its power lay in providing something the people of the Eastern Bloc were starved for: uncensored, objective news and information. It reported on world events truthfully, exposed the lies and corruption of the communist regimes by simply reporting the facts, and played Western music and cultural programs that were officially forbidden.

This was a long-term information war. For decades, RFE/RL served as a symbol of the free world and a constant reminder to those living under tyranny that another reality existed. It systematically undermined the legitimacy of the Soviet-backed governments, fostered dissent by breaking the state's monopoly on information, and kept the hope of freedom alive. Many leaders of the eventual anti-communist revolutions that swept through Eastern Europe in the late 1980s, such as Poland's Lech Wałęsa and the Czech Republic's Václav Havel, later credited RFE/RL with being a vital source of inspiration and a critical tool in the eventual collapse of the Soviet empire.[8] It was a stunning victory achieved not with missiles, but with microphones – a testament to the strategic power of broadcasting the truth into a world of lies.

Spiritual Warfare: The Assault on the Mind

Just as earthly armies battle for the minds of soldiers and civilians, our enemy wages a relentless information war for control of our thoughts. This is his primary theater of operations in the initial assault. He understands that if he can control your thought life, he can eventually control your emotions, your decisions, and your destiny. He launches a constant barrage of spiritual propaganda and psychological operations (PSYOPs) designed to demoralize you, destabilize your emotions, and disrupt your connection with your Commander.

This is not a peripheral struggle; it is the central battle. The enemy attacks our minds with negative thoughts, doubts, anxieties, and fears. He plants seeds of doubt about God's love, His power, and His faithfulness. He whispers insidious lies that sound like our own internal voice, our own rational conclusions. This is the nature of his PSYOPs campaign; it is most effective when the target does not realize he is being manipulated. These targeted lies include:

Lies Targeting Your Identity and Worth: These lies are designed to make you question who you are in Christ and to foster feelings of inadequacy and insecurity.

o **"You are a fraud. If people in your church knew the *real* you – your thoughts, your secret sins – they would reject you completely."** (Objective: Create imposter syndrome and prevent authentic fellowship).

o **"Your past defines you. You can call yourself a 'new creation,' but you'll always be the person who did [X, Y, Z]."** (Objective: To nullify the power of the cross and keep you chained to past shame).

o **"Look at how much more spiritual/gifted/successful that person is. You will never measure up to them."** (Objective: To sow seeds of envy, comparison, and discontentment, distracting you from your own unique calling).

Lies Targeting Your Relationship with God:

These lies are designed to create distance and mistrust between you and your Commander.

o **"God is perpetually disappointed in you. He loves you, but He doesn't *like* you right now."** (Objective: To replace the truth of God's grace with a performance-based relationship, leading to fear and spiritual exhaustion).

o **"That promise in the Bible is for other people – for 'real' Christians, not for someone like you with your struggles."** (Objective: To personalize doubt and convince you that God's Word is not applicable to your specific situation).

o **"God is holding out on you. His rules are designed to keep you from true happiness and fulfillment."** (Objective: An echo of the serpent's lie in the Garden, painting God as a cosmic killjoy).

Lies Targeting Your Sin and Grace:

These are a two-front assault, either minimizing sin to encourage it or exaggerating its hold to create despair.

- ○ (The 'Soft' Lie): **"This 'small' sin isn't a big deal. It's a gray area. God's grace covers it, so you don't need to fight it so hard."** (Objective: To twist the doctrine of grace into a license for compromise, creating a foothold for a larger stronghold).

- ○ (The 'Hard' Lie): **"You've crossed a line this time. You've abused God's grace too often. You are too far gone for real restoration."** (Objective: To induce a spirit of hopelessness and self-condemnation, preventing you from seeking the very forgiveness that is waiting for you).

  - ○ **"You have to 'fix' yourself and prove you're sorry before you can approach God in prayer or go back to church."** (Objective: To promote a works-based righteousness that completely negates the purpose of the throne of grace).

Lies Targeting Your Ministry and Calling:
These lies are designed to neutralize your effectiveness for the Kingdom.

- ○ **"You aren't making any real difference. All your effort is pointless and nobody is changing."** (Objective: To cause burnout, discouragement, and a desire to quit your ministry or service).

- ○ **"If God had truly called you to this, it wouldn't be this difficult. This struggle is proof that you're outside of His will."** (Objective: To equate hardship with disobedience, making you abandon a difficult but God-ordained assignment).

- ○ **"You don't have what it takes. You're not smart enough, charismatic enough, or spiritual enough for this role."** (Objective: To attack your confidence in God's equipping power, causing you to shrink back from your calling).

Lies Targeting Your Relationships and Community:

These lies are designed to achieve the tactical goal of division and isolation.

o **"They are definitely talking about you behind your back. You can't trust them."** (Objective: To sow paranoia and suspicion, preventing the vulnerability required for true fellowship).

o **"That comment they made wasn't a mistake; it was a deliberate attack. They meant to hurt you."** (Objective: To foster a spirit of offense and bitterness, preventing reconciliation and destroying unity).

o **"You're the only one who struggles like this. If you tell anyone, they will judge you or think less of you."** (Objective: To use shame to enforce secrecy and isolation, which is the perfect environment for sin to grow).

These are not random, fleeting thoughts. They are targeted munitions in a psychological war. They are the enemy's "Axis Sally," broadcasting a frequency of despair directly into your mind, designed to erode your spiritual morale and cripple your will to fight. He knows that a believer who is consumed by fear, paralyzed by anxiety, and convinced of their own inadequacy is a believer who has been effectively neutralized on the battlefield without a single shot being fired in the physical realm. His goal is to achieve a psychological victory that makes a spiritual surrender inevitable.

Case Study: Cain and the Unchecked Thought (Genesis 4)

The first recorded murder in history was the direct result of a lost battle in the mind. This encounter serves as the foundational case study for spiritual information warfare. After God accepted Abel's offering but rejected Cain's, Cain became "very angry, and his face was downcast." God Himself intervened at this critical moment, giving Cain a direct intelligence briefing on the psychological battle raging within him and providing a clear path to victory.

The Enemy's Psychological Assault: The initial strike was not an external temptation, but an internal one. The enemy planted thoughts of jealousy, resentment, self-pity, and murderous anger in Cain's mind. These thoughts were the first wave of the assault, designed to destabilize his emotions and push him toward violence. They were the enemy's propaganda, telling him he was a victim and that his anger was justified.

The Divine Counsel (The Counter-PSYOPs Briefing): God gave Cain a clear and actionable counterintelligence strategy: "If you do what is right, will you not be accepted? But if you do not do what is right, sin is crouching at your door; it desires to have you, but you must rule over it" (Genesis 4:7). This was a divine command to engage in mental warfare. God told Cain to take authority over the sinful thought pattern – to "rule over it" – before it escalated into a catastrophic action.

The Failure of Counterintelligence: Cain did not take the thought captive. He nursed it. He allowed the enemy's broadcast of anger and jealousy to occupy the entire airspace of his mind. He failed to rule over the sin crouching at his door, and as a result, the psychological assault escalated into a physical one. He lured his brother into the field and murdered him. Cain lost the war because he first lost the battle in his mind. This demonstrates the critical principle that unchecked thoughts are the gateway to unconquered sins.

Counterintelligence: Taking Every Thought Captive

The only way to win an information war is to seize control of your own mind. You cannot be a passive recipient of every thought that drifts through your consciousness. You must become a vigilant gatekeeper, a disciplined intelligence officer who actively interrogates every thought and challenges its right to be there. An intelligence agency doesn't just defend against enemy propaganda by trying to block it; it actively trains its people to recognize disinformation, to analyze its source and motive, and to reject it based on known truth. It launches a counter-propaganda campaign to ensure its own people are inoculated against the enemy's lies.

The Apostle Paul provides the definitive counterintelligence directive for this mental battle:

"We demolish arguments and every pretension that sets itself up against the knowledge of God, and we **take captive every thought to make it obedient to Christ.**" – 2 Corinthians 10:5

This is an aggressive, proactive command. It is not about simply thinking positive thoughts; it is about waging a counter-offensive against enemy propaganda at the source. This requires a disciplined, multi-step process.

Directive 1: Identify the Enemy Broadcast

You must learn to recognize the enemy's frequency. His broadcasts are always characterized by the fruit of his kingdom: fear, anxiety, condemnation, confusion, hopelessness, and accusation. When a thought enters your mind that produces these feelings, you must immediately identify it as a potential enemy transmission. Do not accept it as your own. Challenge its origin.

A thought that tells you God is disappointed in you is not from the God who says there is "no condemnation for those who are in Christ Jesus" (Romans 8:1). A thought that tells you to be anxious is not from the God who commands, "Do not be anxious about anything" (Philippians 4:6). A thought that tells you that you are a worthless fraud is not from the God who calls you a "new creation" (2 Corinthians 5:17). You must become an expert at recognizing the tone and signature of the enemy's propaganda, marking it as hostile the moment it appears on your mental radar.

Directive 2: Arrest and Interrogate the Thought

Once you have identified a hostile thought, you are commanded to "take it captive." This is the language of a soldier arresting an enemy spy on sight. You stop the thought in its tracks. You refuse to let it run rampant in your mind, and you certainly do not entertain it or offer it a seat at the table. Then, you subject it to interrogation by holding it up against the unwavering truth of God's Word. This is a deliberate, conscious act.

The Lie: "You've failed in this area again. You will never truly be free. This sin will always be your master."

The Interrogation: Is this thought obedient to Christ? No. The truth of Christ is, "Sin shall not be your master, because you are not under law, but under grace" (Romans 6:14). The truth of Christ is, "It is for freedom that Christ has set us free" (Galatians 5:1).

The Lie: "If God really loved you, He wouldn't let you go through this painful trial. He has abandoned you."

The Interrogation: Is this thought obedient to Christ? No. The truth of Christ is that nothing "will be able to separate us from the love of God that is in Christ Jesus our Lord" (Romans 8:39). The truth of Christ is that God "disciplines those he loves" (Hebrews 12:6) and that He works all things together for the good of those who love Him (Romans 8:28).

This process of interrogation exposes the lie by contrasting it with inarguable, documented truth from your Commander.

Directive 3: Launch a Counter-Broadcast of Truth

It is not enough to simply stop a negative thought. You must replace it. This is the essence of renewing your mind (Romans 12:2). You must launch a powerful counter-broadcast, filling your mind with the truth of Scripture until it drowns out the enemy's frequency. This is why memorizing Scripture is so critical; it is the process of stockpiling the ammunition you will use in this fight.

When the enemy broadcasts fear, you counter with 2 Timothy 1:7: "For God has not given us a spirit of fear, but of power and of love and of a sound mind." When he broadcasts accusation, you counter with Romans 8:31: "If God is for us, who can be against us?" When he broadcasts lies about your identity, you counter with 1 Peter 2:9: "But you are a chosen people, a royal priesthood, a holy nation, God's special possession." Speaking these truths, even out loud, is the spiritual equivalent of seizing the enemy's radio tower and broadcasting your own victory message.

Directive 4: Practice Proactive Mental Discipline

Winning the information war requires proactive, daily training. You cannot wait until you are in a full-blown crisis of anxiety to start fighting. You must train your mind daily to dwell on the right things, making it a hostile environment for the enemy's propaganda. Paul gives us the training regimen in Philippians 4:8:

"Finally, brothers and sisters, whatever is true, whatever is noble, whatever is right, whatever is pure, whatever is lovely, whatever is admirable – if anything is excellent or praiseworthy – think about such things."

This is a command to proactively set your mind's frequency to the things of God. By deliberately focusing your attention on truth, goodness, and beauty, you are building powerful mental and spiritual defenses. You are not just reacting to enemy attacks; you are creating a mental fortress so strong and so saturated with the truth of God that the enemy's lies cannot find a place to land.

Biblical Assault:
After-Action Reports on Mental Warfare

The battle for the mind is a central theme of Scripture, providing us with clear case studies of both catastrophic failure and resilient victory. These accounts serve as tactical debriefings, revealing the consequences of allowing enemy propaganda to go unchecked and the power of holding fast to the truth under immense psychological pressure.

Case Study: Cain and the Unchecked Thought (Genesis 4)

The first recorded murder in history was the direct result of a lost battle in the mind. This encounter serves as the foundational case study for spiritual information warfare, demonstrating how an unchallenged thought escalates into a destructive act. After God accepted Abel's offering but rejected Cain's, Cain became "very angry, and his face was downcast." God Himself intervened at this critical moment, giving Cain a direct intelligence briefing on the psychological battle raging within him and providing a clear path to victory.

The Enemy's Psychological Assault: The initial strike was not an external temptation, but an internal one. The enemy planted thoughts of jealousy, resentment, self-pity, and murderous anger in Cain's mind. These thoughts were the first wave of the assault; the enemy's propaganda designed to destabilize Cain's emotions and tell him that his anger was justified and that he was a victim.

The Divine Counsel (The Counter-PSYOPs Briefing): God gave Cain a clear and actionable counterintelligence strategy. He diagnosed the threat: "sin is crouching at your door; it desires to have you." He then issued a direct command for mental warfare: "but you must rule over it" (Genesis 4:7). This was a divine order to engage in the battle of the mind – to take authority over the sinful thought pattern, to arrest it, and to master it before it could escalate.

The Failure of Counterintelligence: Cain did not obey the command. He did not take the thought captive. Instead, he nursed it, meditated on it, and allowed the enemy's broadcast of anger and jealousy to occupy the entire airspace of his mind. He failed to rule over the sin crouching at his door. The result was that the psychological assault metastasized into a physical one. He lured his brother into the field and murdered him. Cain lost the war because he first lost the battle in his mind, providing a permanent lesson on the fatal consequences of unchecked thoughts.

Case Study: Job and the Sustained Assault

The book of Job is a record of one of the most intense and sustained psychological warfare campaigns ever launched against a single individual. After losing his children, his wealth, and his health in a series of catastrophic attacks, Job was subjected to a relentless barrage of demoralizing propaganda from his own friends.

The Enemy's Psychological Assault: Job's friends, operating as unwitting agents of the accuser, launched a campaign to break his spirit. Their core message was a toxic and persistent lie: "You must have sinned. This is your fault. God is punishing you for some hidden transgression." This was a direct assault on Job's integrity and his understanding of God's character. For chapter after chapter, they hammered him with this false narrative, trying to force him into a false confession and drive him into a pit of despair.

The Biblical Assault (Job's Defense): Despite his immense suffering and profound confusion, Job refused to accept their false narrative. His defense was a showpiece in resisting psychological warfare. He maintained his innocence and, critically, he refused to "curse God and die" as his wife suggested. Though he questioned God and wrestled honestly with his pain, he never abandoned his core belief in God's ultimate sovereignty and goodness. Even in his darkest moment, he declared, "Though he slay me, yet will I hope in him" (Job 13:15). He held fast to the truth of his own integrity before God and his belief in God's character, even when all the "evidence" presented by his friends screamed the opposite. Job won the information war by refusing to confess to a lie, even under immense and prolonged psychological pressure. His mind, though battered, was not captured.

Closing Charge: Win the War Within

You have been shown the battlefield where the enemy's primary assault takes place: the battlefield of your mind. He is a master of psychological warfare. He will broadcast lies of fear, accusation, and hopelessness, hoping to defeat you through demoralization long before he needs to engage you in a battle of will. He wants to convince you that his propaganda is your own reality.

But you are not a helpless civilian, forced to listen to the enemy's broadcast. You are a warrior, and you have been given both the authority and the weapons to control the airspace of your own mind. The Word of God is your counter-broadcast of truth. The discipline of taking thoughts captive is your defensive perimeter. The renewing of your mind is your long-term strategy for victory.

Do not be a passive listener to the enemy's lies. Do not allow his frequency of fear and doubt to become the background noise of your life. You have a choice. You can, like Cain, allow a toxic thought to fester until it becomes a destructive action. Or you can, like Job, hold fast to the truth even when your circumstances and the voices around you scream otherwise.

The war for your spiritual life will be won or lost in the six inches between your ears. Take command of your thoughts. Challenge the lies. Demolish the arguments that set themselves up against the knowledge of God. Saturate your mind with the truth of who God is and who you are in Him.

Win the war within, and you will be unbeatable on the outside.

Chapter 9

Suppression of Enemy Defenses (Silencing the Truth)

Military Warfare: Kicking in the Door

AFTER ACHIEVING CONTROL of the skies, the next critical step in a modern assault is the Suppression of Enemy Air Defenses, or SEAD.[1] This is a specialized and violent operation designed to blind, deafen, and neutralize the enemy's ability to fight back against the main air offensive. It is the act of "kicking in the door" so that the primary assault force – the heavy bombers and ground-attack aircraft – can strike their targets without being shot down. A successful SEAD campaign creates a safe corridor through the most dangerous enemy territory, rendering their defensive weapons useless.

The targets of a SEAD campaign are specific and vital: enemy radar installations that detect incoming aircraft, command-and-control bunkers that coordinate defensive actions, and the surface-to-air missile (SAM) batteries and anti-aircraft artillery (AAA) that pose a direct threat. The methods are a symphony of coordinated violence. Stealth aircraft, like the F-117 Nighthawk, slip past enemy radar to drop precision bombs on key command centers.[2] Cruise missiles, launched from ships hundreds of miles away, fly low to the ground to destroy known SAM sites. And specialized "Wild Weasel" aircraft fly daring missions, intentionally baiting enemy radar to lock onto them, only to fire a high-speed anti-radiation missile back down the radar beam, destroying the source.[3] The goal is to systematically dismantle the enemy's entire defensive network, leaving them exposed and vulnerable to the overwhelming force that is to follow.

Case Study: The Opening Night of Operation Desert Storm (1991)

The SEAD campaign that opened the Gulf War was a prime example of modern electronic and kinetic warfare. Before the main waves of bombers and strike fighters crossed into Iraq, a specialized force was sent to dismantle Saddam Hussein's sophisticated air defense system. A team of Apache helicopters, flying deep into enemy territory, destroyed key early-warning radar sites, creating a 20-mile-wide blind spot in the Iraqi radar network.[4]

Simultaneously, U.S. Navy warships in the Persian Gulf launched a barrage of Tomahawk cruise missiles aimed at command headquarters and critical defense nodes in Baghdad.[5] Then, the F-117 Nighthawk stealth fighters, invisible to what remained of the Iraqi radar, slipped into the capital and began systematically destroying the most important strategic targets.[6] By the time the main, non-stealthy assault force arrived, the Iraqi air defense network was in a state of chaos. Its eyes were blinded, its command structure was shattered, and its ability to effectively resist the air campaign was crippled. This initial, overwhelming suppression of defenses ensured that the Coalition would own the skies for the rest of the war with minimal losses.

Case Study: Operation Mole Cricket 19 (1982)

During the 1982 Lebanon War, the Israeli Air Force (IAF) executed one of the most devastatingly successful SEAD operations in history against the Syrian SAM missile batteries stationed in Lebanon's Bekaa Valley. The Syrians had created a dense and formidable air defense network that they believed was nearly impenetrable. The IAF, however, had meticulously studied the system and prepared a comprehensive plan to destroy it.

In a single afternoon, the IAF launched a massive, coordinated attack. They used drones to act as decoys, tricking the Syrians into turning on their powerful radar systems. The moment the radars were activated, Israeli fighters launched a wave of anti-radiation missiles that homed in on the signals, destroying the radar arrays. Simultaneously, other strike fighters, guided by real-time intelligence from scout planes, swooped in to destroy the missile launchers themselves. In the span of about two hours, the IAF had completely obliterated 19 Syrian SAM batteries with almost no losses to their own forces.[7] It was a stunning demonstration of how a well-planned and ruthlessly executed SEAD campaign can dismantle even the most formidable defenses.

Spiritual Warfare: Suppressing the Gospel

Just as a modern military force must suppress enemy defenses before its main attack can succeed, our adversary's primary objective in the initial assault is to suppress the one weapon he cannot defeat: the truth of the Gospel. He knows that the message of Jesus Christ – His life, death, and resurrection – is the "power of God that brings salvation to everyone who believes" (Romans 1:16). It is the divine "bunker buster" that can demolish any stronghold in the human heart, the ultimate truth that liberates a soul from his kingdom of darkness. Therefore, he dedicates enormous strategic effort to silencing, discrediting, and neutralizing the proclamation of this truth in the world.

This is Satan's global SEAD campaign. His goal is to create a cultural and intellectual environment where the Gospel is "jammed," where its signal is suppressed, and where those who proclaim it are neutralized before their message can hit its target. He does this not by trying to disprove the truth – he cannot – but by making the world deaf to its sound and hostile to its messengers. His methods are pervasive and strategic:

Attacking the Source: He relentlessly attacks the credibility and authority of the Bible. He promotes the narrative in academia and popular culture that Scripture is an unreliable, outdated, and contradictory collection of myths written by ignorant men. He wants to suppress the Gospel by convincing the world that its source document is fundamentally untrustworthy.

Promoting Alternative Spiritualities: He floods the spiritual marketplace with a thousand alternative gospels, from New Age philosophies and Eastern mysticism to the gospel of secular self-help and political ideologies. These are his decoy targets, designed to draw seeking souls away from the one, exclusive truth of salvation in Christ. He creates a noisy, confusing spiritual landscape where the unique signal of the Gospel is lost in the static.

Intimidation and Persecution: In many parts of the world, this suppression is overt and violent. Believers are imprisoned, tortured, and killed for proclaiming their faith. But in the West, the persecution is often more subtle, a psychological SEAD campaign. It comes in the form of social ridicule, professional marginalization, and the cultural stigma of being labeled as hateful, intolerant, or bigoted for holding to biblical truth. He seeks to silence believers by making the personal cost of speaking the truth too high to bear.

Inducing Self-Censorship: Perhaps his most effective tactic is to make believers suppress themselves. He infiltrates our own minds, making us feel ashamed or embarrassed to share our faith. He whispers the strategic lies that our faith is a "private matter," that we shouldn't "force our beliefs on others," or that we are not equipped enough to answer the hard questions. He wants to convince the soldier to keep his sword in its sheath out of a fear of looking foolish.

As the Apostle Paul stated with chilling clarity, the "god of this age has blinded the minds of unbelievers, so that they cannot see the light of the gospel that displays the glory of Christ" (2 Corinthians 4:4). This blinding is an active, ongoing SEAD campaign, designed to ensure the most powerful weapon in God's arsenal never leaves the silo.

Case Study: The Parable of the Sower (Matthew 13)

In this parable, Jesus Himself provides the ultimate intelligence briefing on the enemy's SEAD tactics. A farmer sows seed, which represents the "word of the kingdom." The seed is perfect, but it falls on four different types of soil, a
nd in three of those cases, the enemy successfully suppresses its effectiveness.

The Hard Path (Direct Interception): Some seed falls on the path, and "the birds came and ate it up." Jesus explains this plainly: "When anyone hears the message about the kingdom and does not understand it, the evil one comes and snatches away what was sown in their heart." This is a direct, immediate suppression. The enemy doesn't even allow the seed to germinate. He snatches the truth away before it can be understood or considered, often using prejudice, cynicism, or immediate distraction.

The Rocky Ground (Suppression by Persecution): Some seed falls on rocky ground, where it springs up quickly but has no root. "When trouble or persecution comes because of the word, they quickly fall away." This is suppression through intimidation. The initial joy of the message is suppressed by the heat of social pressure, ridicule, or overt opposition. The hearer decides the cost of holding onto the truth is too high, and the Gospel is abandoned.

The Thorny Ground (Suppression by Distraction): Some seed falls among thorns, which "grew up and choked the plants." Jesus identifies these thorns as the "worries of this life and the deceitfulness of wealth." This is suppression by strategic distraction. The Gospel is not rejected outright; it is simply choked out by competing priorities. The pursuit of wealth, career, pleasure, and worldly security grows more vigorously, crowding out the truth until it becomes fruitless and spiritually dead.

In this single parable, Jesus exposes the enemy's entire SEAD playbook: direct interception of the message, suppression of the messenger through persecution, and the choking of the message through the noise of worldly distractions.

Counterintelligence: Refusing to Be Silenced

When faced with an enemy campaign designed to suppress your primary weapon system, the only effective counter strategy is to refuse to be silenced. A soldier does not lay down his rifle because the enemy is shooting at him; he learns to shoot back more effectively and from a more defensible position. A pilot does not ground his aircraft because of enemy anti-aircraft fire; his command redoubles its efforts to clear a safe corridor. Our counterintelligence against the enemy's global SEAD campaign is to become more bold, more wise, and more relentless in the proclamation of the Gospel. It is a deliberate choice to transmit the truth on every frequency, regardless of the enemy's attempts to jam the signal.

The enemy's strategy is sophisticated. He will not always meet you with a direct frontal assault. He will use the subtle pressures of culture, the intellectual arrogance of academia, and the social fear of ostracism to convince you that the best course of action is to stand down, to privatize your faith, and to remain silent. Our counter-maneuvers, therefore, must be equally strategic, fortifying our resolve and preparing us for effective engagement.

Directive 1: Know Your Weapon (Master the Gospel)

You cannot boldly and effectively deploy a weapon you do not understand. A soldier must be able to strip down and reassemble his rifle in the dark. He must know its capabilities, its limitations, and have unshakeable confidence in its reliability. To counter the enemy's suppression of the Gospel, you must first have a deep, personal, and unshakeable confidence in its truth and power. This goes far beyond knowing a simple evangelistic tract.

It means understanding the theological richness of what Christ accomplished on the cross. It means studying the doctrine of the atonement – that Christ's death was a substitutionary sacrifice that satisfied the wrath of God against sin. It means meditating on the historical and spiritual reality of the resurrection – the event that validates every claim Jesus ever made. You must grasp that the Gospel is not merely a fire escape from hell; it is the very power of God to transform a human life from the inside out (Romans 1:16). A believer who is awestruck by the depth, power, and beauty of the Gospel is one who will be eager to share it, not ashamed of it.

Directive 2: Prepare for Engagement (Apologetics and Testimony)

A significant part of the enemy's suppression strategy is to make you fear you won't be able to answer the hard questions. He wants you to feel intellectually cornered and foolish. The counter to this is preparation. The Apostle Peter gives us a clear command: "But in your hearts revere Christ as Lord. Always be prepared to give an answer to everyone who asks you to give the reason for the hope that you have. But do this with gentleness and respect" (1 Peter 3:15).

This preparation is twofold:

1. **Intellectual Preparation (Apologetics):** You do not need to be a seminary professor, but you should take the time to learn the basics of defending your faith. Learn the fundamental historical evidence for the resurrection. Understand the common logical fallacies used in arguments against Christianity. Familiarize yourself with how to answer basic questions about the reliability of the Bible or the problem of evil. This preparation removes the fear of the unknown and equips you to engage in conversations with confidence.

2. **Personal Preparation (Your Testimony):** Even more powerful than an intellectual argument is your personal story. This is the one piece of evidence the enemy cannot refute. Prepare your testimony. Practice articulating it clearly and concisely. Be able to explain what your life was like before Christ, how you came to know Him, and the specific, tangible difference He has made in your life. This is your unique, powerful, and undeniable contribution to the battle.

Directive 3: Live an Unimpeachable Life
(The Power of a Credible Witness)

The enemy loves nothing more than to silence the Gospel by pointing to the hypocrisy of the messenger. He will use your personal failures, your anger, your greed, or your impatience to discredit your public message. He will whisper to the world, "Why would you listen to them? Their lives are no different than yours."

Therefore, one of the most powerful forms of counterintelligence is to live a life of integrity. When your actions align with your words, your testimony becomes powerfully credible. A life increasingly characterized by the fruit of the Spirit – love, joy, peace, patience, kindness, goodness, faithfulness, gentleness, and self-control (Galatians 5:22-23) – is a direct assault on the enemy's narrative that Christianity is oppressive and hypocritical. It doesn't mean you must be perfect, but it means you must be authentic, quick to repent, and genuinely striving to live a life that honors Christ. Your character becomes the unimpeachable verification of your message.

Directive 4: Embrace Strategic Boldness

To counter a campaign of silence and intimidation, you must be bold. This does not mean being obnoxious, argumentative, or needlessly provocative. Strategic boldness is refusing to be intimidated into self-censorship. It means looking for natural opportunities to turn a conversation toward spiritual matters. It means lovingly sharing the reason for your hope when asked. It means standing firm for biblical truth in your workplace or community, even when it is unpopular.

As Jesus told us, we are blessed when we are insulted and persecuted for His sake (Matthew 5:11-12). This blessing is the Commander's assurance that we are taking effective fire. It means our proclamation of the truth is hitting its target and disrupting the enemy's plans. The fear of what others might think is a key part of the enemy's SEAD campaign. Strategic boldness is the act of pushing through that fear, trusting that the power is in the message, not the messenger, and leaving the results to God.

Biblical Assault:
The Early Church Refuses to be Silenced

The Book of Acts gives us the clearest picture of how the enemy attempted to suppress the truth of the Gospel – and how God's people responded. From the moment the Holy Spirit empowered the disciples at Pentecost, the proclamation of Christ crucified and risen became the unstoppable weapon of the Church.

In Acts 4, Peter and John were arrested after healing a lame man and preaching Jesus in the temple. The rulers, elders, and teachers of the law recognized the danger of this message: "They commanded them not to speak or teach at all in the name of Jesus" (Acts 4:18). This was a classic SEAD maneuver — an attempt to intimidate the messengers and suppress the Gospel before it could spread further.

But the apostles' response was the counterintelligence of bold faith: "Which is right in God's eyes: to listen to you, or to him? You be the judges! As for us, we cannot help speaking about what we have seen and heard" (Acts 4:19–20). The very attempt to silence them only confirmed the necessity of their mission.

Later, in Acts 5, the apostles were again arrested and ordered not to speak in the name of Jesus. This time the stakes escalated, as they were flogged for their obedience. Yet their reaction was astonishing: "The apostles left the Sanhedrin, rejoicing because they had been counted worthy of suffering disgrace for the Name. Day after day, in the temple courts and from house to house, they never stopped teaching and proclaiming the good news that Jesus is the Messiah" (Acts 5:41–42).

The biblical record makes clear that the Gospel has always been met with suppression campaigns — whether through intimidation, violence, or ridicule. Yet it also testifies that every attempt to silence the message only amplified it. Persecution scattered the Church, but in scattering, it spread the Gospel farther (Acts 8:1, 4). The harder the enemy pressed, the more the truth of Christ multiplied.

The lesson for us is simple and direct: the only way the enemy's suppression campaign succeeds is if God's people willingly choose silence. The biblical assault is not an abstract encouragement; it is a battle plan written in the blood of the early Church. To follow Christ is to pick up this same weapon — the proclamation of His name — and refuse to lay it down, regardless of cost.

Closing Charge: The Unsilenceable Message

You have seen the enemy's initial assault doctrine: the suppression of your primary weapon. He will try to silence you. He will use the culture to make you feel irrelevant, he will use shame to make you feel hypocritical, and he will use fear to make you feel unsafe. He wants you to believe that your voice does not matter, that your faith is a private affair, and that the truth of the Gospel is a message to be guarded quietly rather than proclaimed boldly. He seeks to neutralize the Church by convincing its warriors to keep their swords in their sheaths.

But you have been entrusted with an unsilenceable message. The Word of God cannot be burned by kings. The testimony of the resurrection cannot be contained by threats. The truth of the Gospel is the most powerful force in the universe, and it has been placed in your hands. To be silent is to cooperate with the enemy's strategy. To speak is to engage in a divine assault.

Therefore, your charge is to refuse to be suppressed.

Know the Message. Be so confident in the Gospel that you are eager to share it. Understand its power, and you will never be ashamed of it.

Prepare to Engage. Do not be caught off guard. Be ready to give a reason for the hope that you have, both with logical answers and with your personal story.

Live a Credible Life. Let your actions be the powerful confirmation of your words. Integrity is a weapon that disarms the accuser and validates your message.

Embrace Boldness. Do not mistake politeness for piety. You are called to speak the truth in love, but you are, above all, called to speak it.

The world is not waiting for another opinion; it is dying for an encounter with the truth. You are the messenger. Will you allow the enemy's suppression campaign to succeed? Or will you, like Peter and John, declare that you cannot help but speak of the great things you have seen and heard?

Your voice is a weapon.

It is time to **deploy** it.

Chapter 10

The Initial Offensive (Exploiting Our Weaknesses)

Military Warfare: The Lightning Strike

IN THE INITIAL phase of an assault, after the battle for the skies has begun and the enemy's defenses are being suppressed, a commander will often seek a swift, decisive victory through a rapid, powerful offensive. This is not a broad, grinding advance along a wide front; it is a concentrated, lightning-fast strike aimed at a single, critical vulnerability in the enemy's lines. The goal of this initial offensive is to achieve "shock and awe" – to shatter the enemy's psychological balance, disrupt their command and control, and seize a key objective so quickly that they are unable to mount an effective defense.[1] This type of operation is designed to bypass the enemy's strengths and strike directly at their weaknesses, leveraging speed and surprise to achieve a disproportionately devastating effect.

These rapid deployments can take many forms: a massive armored thrust punching through a weakly defended sector, an amphibious landing on an unexpected part of a coastline, or an air assault to seize a vital piece of terrain deep behind enemy lines.

Whatever the method, the principle is the same: avoid a costly head-on collision with the enemy's main force. Instead, find the gap, the seam, the unguarded flank, and pour overwhelming force through it before the enemy commander can even understand what is happening. A successful initial offensive can paralyze an entire army, rendering its main strength irrelevant because its foundation has been shattered.

Case Study: The German Blitzkrieg (1940)

The German invasion of France in May 1940 is the archetypal example of a successful initial offensive that exploited a critical enemy weakness.[2] The French and British Allies expected any German attack to be a repeat of World War I – a direct assault through the flat plains of Belgium. To counter this, they had constructed the Maginot Line, a massive and supposedly impenetrable series of fortifications along the French-German border and had positioned their best mobile forces in Belgium.

The German plan, however, was one of brilliant and audacious misdirection. While a secondary force did attack into Belgium to draw the main Allied armies north, the primary German armored thrust – the *Schwerpunkt* (shvair-poongkt) or "focal point" – was aimed at the Ardennes Forest. This was a hilly, densely wooded region that the French High Command considered impassable for a large, armored force.[3] It was their great, un-defended vulnerability.

In a stunning display of speed and coordination, hundreds of German Panzer tanks surged through the Ardennes, supported by Stuka dive-bombers acting as flying artillery.[4] They completely bypassed the Maginot Line and the main Allied armies. Within days, they had crossed the Meuse River and were racing across northern France toward the English Channel.[5] The Allied forces in Belgium were cut off from their supply lines and their chain of command was thrown into chaos. The speed and shock of the German offensive shattered the Allies' psychological will to resist. The German victory was not achieved by defeating the Allied army in a head-on battle, but by striking a single, devastating blow at its most vulnerable point, making the rest of the battle a foregone conclusion.

Case Study: The Capture of Fort Eben-Emael (1940)

On the very first day of the 1940 Blitzkrieg, a small, elite force of German airborne troops executed one of the most audacious and precise initial offensives in history. Fort Eben-Emael in Belgium was considered the strongest fortress in the world, a massive concrete and steel behemoth that guarded key bridges over the Albert Canal.[6] It was armed with powerful artillery cannons in heavily armored turrets and garrisoned by over 1,200 Belgian soldiers. A conventional ground assault would have been a bloody and prolonged siege.

The Germans, however, had identified a critical vulnerability: the fortress's large, flat roof was virtually undefended against an attack from directly above. In the pre-dawn hours of May 10, 1940, fewer than 90 German paratroopers, flying in silent gliders, landed directly on top of the fortress.[7] They had been trained for months on a full-scale replica. Within minutes, they used revolutionary "shaped charges" – explosives designed to focus their blast – to destroy the observation domes and disable the gun turrets. They effectively blinded and disarmed the fortress from the outside.[8] The Belgian garrison, trapped below in their concrete bunkers and completely bewildered by the sudden, silent attack, were unable to fight back effectively. By the time Belgian reinforcements could arrive, the German ground army had crossed the now undefended bridges. A handful of elite soldiers, deployed with lightning speed against a single, overlooked weakness, had neutralized a fortress that was supposed to hold out for weeks.

These operations demonstrate the core principle of the initial offensive: it is not about the application of brute force, but about the surgical application of force against a critical vulnerability to achieve a swift, paralyzing, and decisive result.

Spiritual Warfare: The Lightning Strike

Just as a general seeks a breakthrough with a lightning strike against a weak point, our enemy often initiates his assault not with a broad, slow-moving temptation, but with a rapid, targeted, and overwhelming spiritual offensive. This is not a siege; this is a Blitzkrieg. It is a sudden, shocking attack aimed directly at the most vulnerable, unguarded sector of our lives – our personal "Ardennes Forest" where our defenses are weakest. The objective of this initial offensive is to achieve a quick, devastating victory, to establish a sinful pattern or an emotional stronghold so quickly that our spiritual equilibrium is shattered before we can mount an effective defense.

This strategy is the culmination of the enemy's pre-conflict preparation. His reconnaissance has identified your specific weaknesses. He has studied your past traumas, your recurring sins, your emotional triggers, and your unguarded moments. He is not guessing where to strike; he is launching a precision-guided munition at a pre-selected target. The Bible makes it clear that this is a personalized assault: "each person is tempted when they are dragged away by their own evil desire and enticed" (James 1:14). The enemy knows your specific desires, and his initial offensive is designed to exploit them with maximum speed and shock value.

This is why these temptations often feel so sudden and overwhelming:

For the person with a temper, it is the one specific comment from a spouse or coworker that bypasses all reason and triggers an immediate, explosive rage.

For the person struggling with lust, it is the unexpected image that flashes on a screen, designed to create an instantaneous and powerful wave of desire before moral defenses can be raised.

For the person prone to despair, it is the one piece of bad news – a medical report, a financial setback – that connects with a past trauma and plunges them immediately into hopelessness, bypassing faith entirely.

For the person with an addiction, it is the sudden, out-of-nowhere craving that feels like an ambush, demanding to be satisfied *now*.

In each case, the strategy is the same: speed and surprise. The attack is designed to be so fast and so tailored to our specific weakness that it bypasses our rational, spiritual mind and appeals directly to the raw, unredeemed desires of our flesh. The enemy seeks to win the battle in seconds, before we even realize we are in a fight.

Case Study: The Fall of David (2 Samuel 11)

The story of David and Bathsheba is the ultimate biblical case study of a successful rapid deployment offensive. In Chapter 1, we examined this event from the perspective of reconnaissance – the enemy studying David's idleness. Now, we will analyze the attack itself.

David was a mighty warrior, a man of incredible courage, and a leader of immense faith. A frontal assault on his courage would have failed. An attack on his leadership would have been repelled. But the enemy, having done his reconnaissance, had identified a critical vulnerability. David, the great king, was idle in Jerusalem while his armies were at war. He was bored, restless, and unaccountable. This was the undefended sector.

The Surgical Strike: The attack was not a complex theological argument or a long negotiation. It was a single visual: a beautiful woman bathing on a nearby rooftop. This was the enemy's "air assault," landing a powerful temptation directly into David's most vulnerable area at a moment when his defenses were at their lowest.

The Rapid Escalation: The offensive unfolded with lightning speed. The sequence was brutally efficient: the **glance**

became a **gaze.** The gaze ignited **desire.** The desire led to an **inquiry.** The inquiry led to a **summons.** The summons led to **adultery.** There was no prolonged siege; it was a rapid, cascading failure. The enemy exploited David's weakness for sensual pleasure and his abuse of royal power so quickly that his spiritual senses were overwhelmed.

The Strategic Result: By exploiting one specific weakness with a rapid, targeted temptation, the enemy neutralized one of the greatest men in biblical history. The initial victory of adultery then led to a campaign of deception and murder, crippling David's moral authority and bringing immense suffering upon his family and his kingdom for years to come. It all began with a single, perfectly timed strike against an unguarded flank.

Counterintelligence: Hardening the Target

To defend against a Blitzkrieg, a lightning-fast armored thrust, an army cannot rely on a thin, evenly defended front line. It must identify its own potential weak points – the "impassable" forests, the unguarded bridges – and proactively fortify them. It must position reserves, lay anti-tank mines, and pre-plan counter attacks. In short, it must "harden the target" *before* the assault begins. Counterintelligence against a rapid deployment offensive is not about reacting in the moment of shock; it is about the disciplined, preemptive work that makes the shock ineffective.

If the enemy's initial offensive is a surgical strike against your most personal and potent weaknesses, then your counterintelligence must be a ruthless campaign of self-assessment and proactive defense. You must become the commander of your own soul, identifying your vulnerabilities with the unflinching eye of a staff officer and establishing defensive protocols long before the first shot is fired.

Directive 1: Proactive Reconnaissance
(Mapping Your Own Ardennes Forest)

A commander who is ignorant of the weak points in his own line is already defeated. Before you can defend against the enemy's exploitation of your weaknesses, you must first identify them yourself with brutal honesty. This is the first and most critical act of personal counterintelligence. You must ask the hard questions:

What is my recurring sin? What is the one temptation that has a consistent track record of defeating me? Be specific. Name it.

What are my emotional triggers? What specific situations, words, or people consistently provoke an ungodly response in me – be it anger, fear, lust, or despair?

What are my points of vulnerability? Am I most vulnerable when I am tired? Lonely? Hungry? Stressed? After a great success (pride)? After a significant failure (shame)?

What past traumas have not been fully surrendered to God for healing? Unhealed wounds are the open gates through which the enemy loves to launch his attacks.

This self-assessment is not an exercise in self-condemnation. It is a strategic necessity. You are identifying the terrain where the enemy is most likely to attack so you can concentrate your defenses there.

Directive 2: Pre-Planned Defenses
(Laying the Minefield)

Once you have identified a vulnerability, you cannot simply hope you'll be strong enough in the moment. You must establish pre-planned, automatic defenses. You must lay a spiritual minefield for the enemy.

Memorize Specific Ammunition: This is more than general Bible reading. If your weakness is anger, you must have verses on patience and self-control memorized and ready to deploy. If it is fear, you must have verses on God's sovereignty and faithfulness chambered and ready to fire. You are pre-selecting the specific weapon for the specific threat you know is coming.

Establish "Rules of Engagement": For your known weaknesses, create simple, non-negotiable rules of engagement for yourself. For example: "If I find myself beginning to gossip, I will immediately excuse myself from the conversation." "If I feel the pull of lust while online, my rule is to immediately shut the device down." "If my spouse says the specific thing that triggers my rage, my rule is to state, 'I need five minutes,' and walk away to pray." These are pre-made decisions that remove the need for willpower in a moment of emotional compromise.

Remove the Means of Attack: A wise commander destroys the bridge the enemy needs to cross. If your weakness is online temptation, installing accountability software is not a sign of weakness; it is a brilliant defensive maneuver. If your weakness is a relationship that consistently leads you into sin, creating distance is not unkind; it is a strategic necessity for survival. Harden the target by making it as difficult as possible for the enemy to exploit your weakness.

Directive 3: The "Battle Drill"
(Immediate Action in a Crisis)

When an ambush is sprung, soldiers do not have time to think; they react based on their battle drills. They do what they have been trained to do, over and over again. We must have our own spiritual battle drills for when a sudden, overwhelming temptation strikes.

The Drill of "Flee": In the face of certain temptations, especially sexual ones, the Bible's command is not to stand and fight, but to *flee* (1 Corinthians 6:18, 2 Timothy 2:22). This is a pre-planned tactical retreat. It is the wisest and most courageous course of action, an immediate disengagement to a more defensible position.

The Drill of "Cry Out": Your immediate action should be to call for fire support. This may not be a long, eloquent prayer. It is often a desperate, one-sentence cry for help: *"God, help me NOW!" "Jesus, save me from this!"* This is the spiritual panic button that alerts High Command that you are in crisis.

The Drill of "Contact": You must have a pre-designated accountability partner – your fire team leader – whom you can contact at a moment's notice. A quick text message saying "Pray for me, I'm under attack" can be the reinforcement that breaks the back of the enemy's assault.

These drills must be pre-decided and rehearsed in your mind. You don't decide to flee in the moment of temptation; you have already decided that fleeing is your standing order.

Directive 4: The After-Action Report (Learning from Engagements)

After the initial assault has passed – whether you stood firm or you fell – the battle is not over. You must conduct an after-action report. Analyze the engagement. If you fell, why? Were you tired? Were you spiritually unprepared? Did you neglect your pre-planned defenses? If you stood firm, why? What worked? Did you deploy your scripture? Did you contact your ally? Every engagement, even a failure, provides critical intelligence that you can use to strengthen your defenses for the next, inevitable attack.

Biblical Assault: After-Action Reports on the Lightning Strike

The enemy's strategy of launching a rapid, concentrated offensive against a known weakness is a recurring theme in Scripture. These accounts serve as critical debriefings, showing us how a moment of unpreparedness, a single unguarded flank, or a specific, unaddressed character flaw can be exploited by the enemy to achieve a swift and devastating victory. By analyzing these campaigns, we learn to recognize the signature of a spiritual Blitzkrieg and the vital importance of hardening our defenses before the assault begins.

Case Study: David and the Rooftop Offensive
(2 Samuel 11)

The story of David and Bathsheba is the quintessential biblical case study of a successful rapid deployment offensive. In Chapter 1, we examined this event from the perspective of reconnaissance – how the enemy studied David's idleness and vulnerability. Now, we will analyze the attack itself, the execution of that intelligence.

David was a mighty warrior, a man of incredible courage, and a leader of immense faith. A frontal assault on his courage would have failed. An attack on his leadership would have been repelled. But the enemy, having done his reconnaissance, had identified a critical vulnerability. David, the great king, was idle in Jerusalem while his armies were at war. He was bored, restless, and – most importantly – unaccountable. This was his personal "Ardennes Forest," the undefended sector of his soul.

The Surgical Strike: The attack was not a complex theological argument or a long negotiation. It was a single visual: a beautiful woman bathing on a nearby rooftop. This was the enemy's "air assault," landing a powerful temptation directly into David's most vulnerable area at a moment when his moral and spiritual defenses were at their lowest. The temptation was perfectly tailored to exploit a latent weakness for sensual pleasure.

The Rapid Escalation: The offensive unfolded with lightning speed. The sequence was brutally efficient: the **glance** became a **gaze.** The gaze ignited **desire.** The desire led to an **inquiry.** The inquiry led to an **abuse of power** (the summons). The abuse of power led to **adultery.** There was no prolonged siege; it was a rapid, cascading failure. The enemy exploited David's weakness so quickly and powerfully that his spiritual senses were overwhelmed. He went from a king at rest to an adulterer in a matter of hours.

The Strategic Result: By exploiting one specific weakness with a rapid, targeted temptation, the enemy neutralized one of the greatest men in biblical history. The initial victory of adultery then forced David into a secondary campaign of deception and murder to cover his tracks, crippling his moral authority and bringing immense suffering upon his family and his kingdom for years to come. It all began with a single, perfectly timed strike against an unguarded flank.

Case Study: Achan and the Spoils of War (Joshua 7)

The story of Achan provides a different but equally powerful example of a rapid offensive, this time exploiting greed in a moment of collective victory. After the miraculous fall of Jericho, God gave a clear and absolute command: all the spoils of the city were "devoted to the LORD" and were to be destroyed or placed in the Lord's treasury. Taking anything for personal gain was strictly forbidden.

The Surgical Strike: In the chaotic aftermath of the victory, while surveying the spoils, Achan was hit with a rapid-fire temptation. He describes the lightning-fast sequence himself: "When **I saw** in the plunder a beautiful robe from Babylonia, two hundred shekels of silver and a bar of gold weighing fifty shekels, **I coveted** them and **took** them" (Joshua 7:21). It was a three-step Blitzkrieg against his heart: See. Covet. Take. The enemy launched an opportunistic assault, exploiting the lust of the eyes and the pride of life in a single, decisive moment.

The Unseen Weakness: Achan's weakness was a hidden, personal greed that was likely unknown to his fellow soldiers. He was part of a victorious, unified army, but he harbored a personal vulnerability that the enemy was able to pinpoint and exploit. He saw an opportunity for personal enrichment and, in a moment of weakness, chose to disobey a direct order from his Commander-in-Chief.

The Strategic Consequence: Achan's rapid, personal sin had devastating corporate consequences. Israel's next military operation, a seemingly simple assault on the small town of Ai, ended in a shocking and humiliating defeat. The entire nation was demoralized, and Joshua was left questioning God's presence. It was only after Achan's sin was exposed and dealt with that Israel could once again be victorious. This after-action report proves that a single, successful lightning strike against one individual's weakness can compromise the integrity and combat effectiveness of the entire army.

Closing Charge: Harden the Front Line

You have seen the enemy's doctrine of the lightning strike – his spiritual Blitzkrieg aimed not at your points of strength, but at your single most vulnerable, unguarded flank. He will not waste his energy in a costly frontal assault against your faith if he can achieve a swift, catastrophic breakthrough by exploiting your temper, your lust, your pride, or your fear. His initial offensive is designed to be a rapid, shocking deployment of a tailored temptation to achieve a quick victory before you even realize you are in a fight.

He is counting on you to leave your own "Ardennes Forest" undefended. He is banking on the fact that you, like the French High Command of 1940, have dismissed your own weaknesses as "impassable" or insignificant. He knows that a single, well-placed, overwhelming temptation can shatter your spiritual balance and lead to a cascade of failure.

But you are not defenseless against this Blitzkrieg. An attack that targets a known weakness is an attack that can be anticipated. A strike that relies on surprise is a strike that can be foiled by preparation. You have been given the intelligence; now you must act on it.

Therefore, your charge is to become a master of your own terrain. You must be the commander who walks his own lines, who is brutally honest about the gaps in his own defenses.

Identify Your Weak Points: Do not live in denial. Know with absolute clarity where the enemy is most likely to strike you. Name your recurring sins. Identify your emotional triggers. This is not an act of condemnation; it is the essential reconnaissance required for survival.

Fortify Your Defenses: Do not simply hope for strength in the moment of attack. Proactively lay your spiritual minefields. Memorize the specific Scriptures that counter your specific weaknesses. Establish non-negotiable rules of engagement for yourself when you feel a trigger being pulled. Make your weak points the most heavily defended sectors of your soul.

Drill Your Immediate Actions: Rehearse your response to a sudden ambush until it is instinct. Know when to flee. Know how to cry out to God for immediate fire support. Have your accountability partner on speed-dial. A pre-planned battle drill is what saves you when there is no time to think.

Will you be a David, caught idle on the rooftop, unprepared for the swift, surgical strike against his unguarded desire? Will you be an Achan, overcome in a moment of victory by a sudden temptation of greed? Or will you be a warrior who has already identified the weak points in your own line and fortified them with the truth of God and the discipline of a soldier?

The lightning strike is coming. But a hardened target can repel even the most powerful assault.

Harden the line.

Shane Cunningham

Chapter 11

Naval Operations – Sea Control (Infiltrating Our Hearts)

Military Warfare: The Highways of Power

BEYOND THE IMMEDIATE clashes on land and in the air, there is another, more fluid theater of war: the sea. The world's oceans are the great highways of global power, the arteries of trade, and the essential medium for projecting force across continents.[1] The nation that achieves "sea control" – the ability to use the sea for its own purposes and to deny its use to the enemy – holds a decisive strategic advantage. Naval operations are not merely about ship-to-ship combat; they are about dominating this vital environment to sustain a war effort and strangle the enemy's ability to fight back.

Establishing sea control is fundamental to all other long-range military operations. It allows a nation to transport and supply its armies, to conduct amphibious landings on hostile shores, and to project its power far from its own borders. Conversely, denying the enemy sea control is a powerful offensive weapon. A successful naval blockade can cut an enemy nation off from critical resources – oil, food, and war materiel – slowly starving its economy and its military machine into submission.[2] The sea lanes are the lifelines of modern nations; controlling them means holding a knife to your enemy's throat.

Case Study: The Battle of Trafalgar (1805)

For years, Napoleon Bonaparte dominated the European continent with his seemingly invincible Grand Army. His one great obstacle to total supremacy was the island nation of Great Britain, protected by the English Channel and the power of the Royal Navy. Napoleon's plan for invasion required him to achieve at least temporary sea control to ferry his army across the channel. The British objective was simple: to ensure that never happened.

The climactic confrontation came at the Battle of Trafalgar. The British fleet, under the command of the brilliant and audacious Admiral Horatio Nelson, engaged the combined fleets of France and Spain.[3] Though outnumbered, Nelson employed a daring and unconventional tactic, breaking the enemy's traditional line of battle and plunging his ships into a chaotic, close-quarters melee where superior British seamanship and gunnery would prove decisive. The result was one of the most complete naval victories in history. The Franco-Spanish fleet was annihilated.[4] While Nelson himself was killed at the moment of his greatest triumph, his victory was absolute. Trafalgar shattered Napoleon's naval power, ended his dream of invading Britain, and established a century of British maritime supremacy. It proved that the master of the land could be defeated by the master of the sea.

Case Study: The Anaconda Plan and the Union Blockade (U.S. Civil War, 1861-1865)

The American Civil War provides a powerful example of how sea control can be used as a long-term strategic weapon of attrition.[5] At the outset of the war, the Union's General-in-Chief, Winfield Scott, proposed a strategy known as the "Anaconda Plan." The plan was to suffocate the Confederacy by sea, blockading its entire coastline from Virginia to Texas, while another army advanced down the Mississippi River, cutting the South in two.

The Union Navy's blockade was a monumental undertaking. Over the course of the war, it grew from a few dozen ships to a fleet of hundreds, tasked with patrolling thousands of miles of coastline.[6] While blockade runners managed to slip through, the naval cordon grew progressively tighter. It starved the Confederacy of imported war materials, industrial goods, and trade revenue.[7] The Southern economy, heavily dependent on cotton exports to Europe, was slowly strangled. The lack of supplies crippled the effectiveness of Confederate armies in the field. The Anaconda Plan was a slow, grinding, and ultimately devastatingly effective use of sea power. It demonstrated that a war can be won not just by winning battles, but by controlling the enemy's arteries of supply and slowly squeezing the life out of their ability to resist.

Spiritual Warfare: Infiltrating the Heart

In the spiritual war, the strategic "sea" is the human heart. It is the deep, often hidden wellspring of our being from which all our actions, attitudes, and words flow. The writer of Proverbs gives us a critical military directive: "Above all else, guard your heart, for everything you do flows from it" (Proverbs 4:23). The heart, in this biblical sense, is not just the seat of emotion; it is the command center of our lives. It is where our deepest affections, loyalties, and desires reside. To achieve "sea control" of the heart is the enemy's ultimate objective in his initial assault.

He knows that if he can infiltrate and control this vital center, he doesn't need to win a hundred smaller battles for our thoughts or actions; he will have captured the source from which they all originate. His strategy is not always a direct, frontal assault of overwhelming temptation. More often, it is a subtle naval campaign of infiltration and blockade.

Infiltration: He seeks to smuggle his agents – pride, selfishness, lust, anger, bitterness, greed – past our defenses and into the harbor of our hearts. Once inside, these agents work to corrupt our affections and turn our loyalties away from God.

Blockade: He works to establish a spiritual blockade, cutting our hearts off from their true source of supply – the grace, peace, and love that flow from the Holy Spirit. He wants to leave us spiritually starved and unable to resist his influence.

An infiltrated heart becomes a base of operations for the enemy. It begins to project his will, not God's, into the world through our words and deeds. A heart controlled by bitterness will produce gossip and division. A heart controlled by lust will produce destructive relationships. A heart controlled by pride will produce selfish ambition. The enemy's naval operation is complete when our lives, flowing from the state of our hearts, begin to serve his purposes.

Case Study: The Corruption of King Saul
(1 Samuel 13, 15, 18)

The life of King Saul is a tragic case study of a heart slowly infiltrated and ultimately captured by the enemy. Saul began as a humble and capable leader, chosen by God. However, a series of strategic infiltrations turned his heart from a vessel of God's will into a fortress of pride, jealousy, and fear.

The first breach came at Gilgal (1 Samuel 13). Pressured by the looming Philistine army and the impatience of his men, Saul disobeyed a direct command from the prophet Samuel and offered a sacrifice himself – a role reserved for the priest. This act was born of fear and impatience, but it opened the door for a more dangerous agent: pride. He had taken spiritual matters into his own hands.

The next infiltration came after his victory over the Amalekites (1 Samuel 15). He was commanded to utterly destroy them, but he spared their king and the best of their livestock, claiming he would sacrifice them to the Lord. Samuel's rebuke was devastating: "To obey is better than sacrifice... For rebellion is like the sin of divination, and arrogance like the evil of idolatry." The agents of pride and self-will were now deeply entrenched in Saul's heart.

The final, decisive infiltration was jealousy. When David defeated Goliath and the people sang, "Saul has slain his thousands, and David his tens of thousands," the Bible says, "Saul was very angry... and from that time on Saul kept a jealous eye on David" (1 Samuel 18:8-9). This jealousy, a powerful agent of the enemy, achieved complete sea control of Saul's heart. It choked out all reason, love, and loyalty. It consumed him, driving him to paranoia, rage, and repeated attempts to murder the very man who served him most faithfully. Saul's heart, once dedicated to God, had been successfully infiltrated and was now a base of operations for the enemy's destructive agenda.

Counterintelligence: Establishing Control of Your Heart

If the enemy's objective is to achieve sea control over your heart, then your counterintelligence must be a deliberate and active campaign to guard that vital territory. You must become the vigilant admiral of your own soul, controlling your ports of entry, conducting internal security sweeps, and ensuring your supply lines to God remain open. This is not a passive or one-time act; it is the moment-by-moment, lifelong work of guarding your heart with all diligence.

Directive 1: Control Your Ports of Entry
(The Eye and Ear Gates)

A naval blockade is useless if the enemy has unguarded ports. Your primary ports of entry are your eyes and your ears. What you consistently watch and listen to will inevitably make its way into the harbor of your heart. You must take active command over what you allow to enter.

Evaluate Your Media Intake: What are you watching? What are you listening to? Does it promote godliness, truth, and love, or does it glorify lust, greed, and cynicism? To leave these ports unguarded is to invite the enemy's contraband into your soul.

Curate Your Relational Environment: The people you spend the most time with will have a profound influence on your heart. Are your closest relationships with people who are pushing you toward Christ, or are they with those who are pulling you into worldliness? Choose your allies wisely.

Directive 2: Conduct Internal Security (Confession and Repentance)

It is not enough to guard the ports; you must also deal with the enemy agents who have already slipped inside. These are the sins, attitudes, and idols that have taken up residence in your heart. This requires the ruthless internal security of honest confession and genuine repentance.

Practice Regular Self-Examination: Ask the Holy Spirit to search your heart and reveal any hidden pride, bitterness, or unholy desire (Psalm 139:23-24). Be willing to see the truth about what is really motivating you.

Be Quick to Confess: When the Spirit convicts you of sin, do not hide it or rationalize it. Confess it immediately and specifically to God. Confession is the act of arresting the enemy's agent and handing him over for judgment.

Repent and Remove: Repentance is more than just saying sorry; it is a decisive turn away from the sin. It involves taking practical steps to remove the idol from the throne of your heart and to starve the sinful desire of its fuel.

Directive 3: Maintain Open Supply Lines (Cultivating the Spirit)

A heart that is cut off from God will become stagnant and vulnerable. You must actively maintain your spiritual supply lines. These are the spiritual disciplines that keep your heart connected to the life-giving power of the Holy Spirit.

Daily Prayer and Worship: This is your primary communication and supply convoy. It is how you receive grace, peace, and direction from your Commander.

Immersion in Scripture: The Word of God is the food that nourishes your heart and the light that exposes the darkness.

Intentional Fellowship: Connecting with other believers provides encouragement and strength, preventing the enemy from isolating you.

Directive 4: Deploy Your Own Fleet (Cultivating the Fruit of the Spirit)

The most effective way to keep the enemy's fleet out of your harbor is to fill it with your own. You go on the offensive by actively cultivating the fruit of the Spirit: love, joy, peace, patience, kindness, goodness, faithfulness, gentleness, and self-control (Galatians 5:22-23). When your heart is actively producing love, there is no room for bitterness. When it is filled with peace, anxiety cannot find a place to anchor. When it is practicing self-control, lust cannot seize the helm. This is not just a defensive posture; it is an offensive occupation of your own heart with the very character of God.

Biblical Assault: After-Action Reports on the Battle for the Heart

The battle for the heart is central to the scriptural narrative, providing clear case studies on both successful defense and catastrophic failure.

Case Study: Joseph and the Siege of Temptation (Genesis 39)

Joseph, a young man serving in the house of Potiphar, found himself the target of a sustained and relentless assault on his heart. Potiphar's wife launched a campaign of seduction, repeatedly demanding that he lie with her. This was a direct attempt to infiltrate his heart with lust and disloyalty.

The Enemy's Assault: The temptation was constant ("day after day"), direct, and came from a position of authority. It offered momentary pleasure at the cost of long-term integrity.

The Biblical Counter Assault: Joseph's defense was a model of successful heart-guarding.

He Acknowledged His Loyalty: His first response was to declare his loyalty to his earthly master, Potiphar, establishing a clear boundary of integrity.

He Identified the True Enemy: He did not see this merely as a human temptation. He elevated the battle to the spiritual realm. His famous declaration, "How then could I do this great wickedness and sin against God?" (Genesis 39:9), reveals that his primary allegiance was to his heavenly Commander. He understood that this was not just about sex; it was about sinning against God.

He Took Physical Action: When the assault became physical, he did not stay and fight a battle of wills. He "fled and got out" (Genesis 39:12). He understood that sometimes the wisest strategy is a tactical retreat to a more defensible position. He chose purity over proximity. Joseph successfully defended the harbor of his heart because he knew who his Commander was and was willing to do whatever it took to remain loyal.

Case Study: Hezekiah and the Infiltration of Pride
(2 Chronicles 32)

King Hezekiah was one of Judah's most righteous kings. He trusted God, and God delivered him miraculously from the Assyrian army and healed him from a deadly illness. But after these great victories, his heart was breached.

The Enemy's Assault: After his healing, the Bible states that "Hezekiah's heart was proud and he did not respond to the kindness shown him" (2 Chronicles 32:25). When envoys from

Babylon came, Hezekiah, in a moment of prideful vanity, gave them a full tour of his palace and his entire treasury, showing off all his wealth and military stores. This was a severe breach of national security, born from a heart infiltrated by pride. He was boasting in his own strength and wealth, not in the God who had given it to him.

The Biblical Counter Assault (Delayed): The Bible gives a chilling analysis: "God left him to test him and to know everything that was in his heart" (2 Chronicles 32:31). The prophet Isaiah was sent to deliver a word of judgment for this act of pride. The good news is how Hezekiah responded. He didn't make excuses like Saul. Instead, "Hezekiah humbled himself for the pride of his heart" (2 Chronicles 32:26). He repented. This act of humility was the counter assault that purged the enemy agent of pride from his heart and delayed the judgment of God. Hezekiah's story serves as a critical warning: even a godly heart can be infiltrated after a great victory, and the only successful counter-maneuver is swift and genuine humility.

Closing Charge: The Captain of Your Heart

You have seen the nature of the naval war. The enemy's objective is to gain control of the strategic seas of your heart, for he who controls the heart controls the life. He will use the subtle tactics of infiltration, smuggling his agents of pride, bitterness, and lust past your defenses. He will attempt to blockade you from the life-giving presence of God, leaving you spiritually isolated and starved for grace.

But you are not a helpless vessel, adrift on a hostile sea. God has made you the captain of your heart, and He has charged you with its defense. The choice is yours. You can leave your ports of entry unguarded, allowing the currents of the culture and the enemy's contraband to pollute your inner harbor. You can, like Saul, allow jealousy and pride to seize the helm and run your life in the ground. Or you can, like Joseph, stand a vigilant watch, refusing to allow wickedness to come aboard. You can, like

Hezekiah, learn to humble yourself, to identify the enemy within, and to throw him overboard.

Guard your heart. Know what you are letting in through your eyes and your ears. Be quick to identify and confess the sins that have already made it past the gates. Actively cultivate the fruit of the Spirit, filling your heart with a fleet so powerful and righteous that there is no room for the enemy's ships.

The battle for your heart is the battle for everything.

Take the **helm**.

Phase III

The Siege

THE INITIAL SHOCK of the assault has faded. The enemy's lightning strikes have been weathered, and his opening offensive has not achieved a quick, decisive victory. Now, the nature of the war changes. The fast-paced battle of maneuver gives way to a new, more grueling phase: **The Siege.**[1]

This is not a war of lightning strikes, but of relentless, grinding attrition. It is a war of patience and pressure, designed to erode your will to fight over the long, hard months and years. In a siege, the attacker surrounds a fortified position, cuts off its supply lines, and systematically works to weaken its defenses until the garrison is either starved into submission or its walls are breached.[2] This is the phase of the war where spiritual burnout happens, where strongholds are established, and where faith is either forged into steel or ground into dust.[3]

In this new phase, the enemy's strategy will escalate. He will take the very tactics he used to prepare the battlefield and turn them into weapons of exhaustion and despair.

In **Phase I,** we learned that the enemy uses **Deterrence** to try and *prevent* you from ever encountering God. Now, in the siege, his objective is more sinister. He will work to **Degrade Your Capabilities,** actively seeking to sever the lines of communication you *already have* with your High Command, leaving you feeling isolated and alone in the battle.

In **Phase I,** we learned that the enemy fights *with* a **Coalition.** Now, in the siege, his primary objective is to *destroy your coalition.* He will work to **Isolate You** from the Body of Christ, cutting you off from all reinforcements and support at the very moment you are under the most intense and sustained pressure.[4]

The chapters in this section will dissect this long war of attrition. We will analyze the enemy's methods for wearing down your spiritual strength, for establishing fortified strongholds of sin in your life, and for suppressing your will to resist through guilt and shame.

Author's Note

I will interrupt the established flow of this manual only once. If this entire book – every strategy, every tactic, every countermeasure – had to be stripped down to a single, life-or-death briefing, it would be this phase of the war: **The Siege.**

The enemy's overt assaults are terrifying. His deceptions are cunning. But they are not where the majority of soldiers fall. The real war, the long war, is won or lost right here, in the grinding, day-in and day-out battle of sheer endurance. This is the phase where Satan uses patience as his most brutal weapon, working to slowly, quietly, and methodically destroy the saints.

He doesn't need to force a dramatic surrender; he only needs to isolate you, erode your beliefs, and make you so utterly weary that you stop resisting, stop caring, and stop getting back up.

The chapters that follow are not just theory; they are your survival guide for the most critical part of the conflict. Pay attention:

This is the **main** battle.

This is the test of **endurance**.

This is the war for your **will**.

Chapter 12

Degrading Enemy Capabilities (Severing the Connection)

Military Warfare: The C2 Attack

WHEN A PROLONGED siege or campaign begins, the attacking force often shifts its focus from destroying enemy combat units to dismantling the very nervous system that allows the enemy to fight. This is the art of attacking Command and Control, or C2.[1] An army, no matter how well-armed or fortified, is rendered impotent if its commander cannot issue orders, if its units cannot coordinate with each other, and if its intelligence cannot be received. The goal of a C2 attack is to degrade the enemy's capability to operate as a coherent force, to sow confusion and paralysis, and to transform a disciplined army into an isolated, ineffective mob.

The targets of a C2 campaign are the critical nodes that enable a military to function.[2] This includes bombing command bunkers and headquarters to eliminate leadership, jamming radio frequencies and cutting fiber-optic cables to sever communication lines, and destroying radar and satellite uplinks to blind the enemy. By degrading these capabilities, a commander can cripple his adversary's ability to react to changing battlefield conditions, to resupply his troops, or to organize a coordinated defense. An army that cannot communicate with its high command is an army that has been orphaned on the battlefield, left to fight and die on its own.

Case Study: The Battle of France (1944)

A powerful and classic example of degrading an enemy's capability to respond can be found in the Allied campaign in France following the D-Day landings in June 1944. After securing a fragile beachhead in Normandy, the greatest single threat to the Allied invasion was the German ability to move their powerful reserve Panzer divisions to the front to drive the invaders back into the sea. The entire success of the operation depended on preventing these elite armored units from arriving in a timely and coordinated fashion.

To achieve this, Allied air forces launched a relentless "interdiction" campaign.[3] This was a massive, sustained effort to degrade Germany's C2 and logistical capabilities throughout France. Day and night, Allied fighter-bombers and heavy bombers targeted the French transportation network. They destroyed key bridges over the Seine and Loire rivers, creating huge geographic barriers. They bombed railway marshalling yards, locomotives, and miles of track, making it impossible for the Germans to move troops and heavy equipment by rail. They strafed roads, ambushing any German convoy that dared to move during daylight hours.

Simultaneously, the Allies activated a powerful coalition on the ground: the French Resistance. Supplied with weapons and explosives, these partisan fighters conducted a widespread campaign of sabotage, cutting telephone lines, derailing trains, and ambushing German officers.[4] The cumulative effect of the air campaign and the Resistance's work was the systematic dismantling of the German nervous system in France. Panzer divisions that should have reached the front in two days took two weeks, arriving piecemeal and having already suffered losses from air attacks.[5] German commanders were often unable to communicate with their superiors or with their own units, leaving them blind, isolated, and paralyzed. The Allies won the battle for Normandy not just by fighting on the beaches, but by ensuring the enemy's most powerful reinforcements were rendered incapable of ever reaching the main battle in time.

Case Study: The Siege of Alesia (52 B.C.)

The ancient world provides an equally powerful, if more brutal, example. When Julius Caesar besieged the Gallic chieftain Vercingetorix at the fortress of Alesia, he faced a two-front problem: the massive Gallic army trapped inside the fortress, and a huge Gallic relief army marching to attack him from the outside. Caesar's genius was in realizing he could defeat both by degrading their capabilities.

He built two massive lines of fortifications. The inner wall, the *circumvallation,* blockaded the fortress, cutting Vercingetorix off from all supplies and reinforcements.[6] The outer wall, the *contravallation,* faced outward and was designed to defend against the Gallic relief army. Caesar's strategy was to degrade the capabilities of both forces simultaneously. Vercingetorix, trapped inside, was slowly starved into submission. The relief army, unable to breach the Roman defenses, could not link up with or resupply their comrades.[7] By isolating the two forces from each other and from their supplies, Caesar rendered both incapable of effective, coordinated action. He defeated them not by overwhelming force, but by a brilliant feat of military engineering that completely degraded their ability to fight.

Spiritual Warfare:
Cutting the Line to Command

In the grinding reality of a spiritual siege, the enemy escalates his strategy beyond the initial assault. He knows that a believer who maintains a strong, clear connection to God can withstand almost any attack. Therefore, his primary objective in a siege is to degrade your spiritual capabilities by launching a direct assault on your line of communication with your High Command. This is a critical escalation from the tactics used in the pre-conflict phase. In **Phase I, his strategy of Deterrence** was designed to *prevent* you from ever approaching God. Now, in the siege, his strategy is more aggressive: he seeks to actively jam, disrupt, and sever the connection you *already have,* leaving you feeling spiritually isolated, deaf to God's voice, and alone in the fight.

His goal is to induce **spiritual apathy** – a state of lukewarmness where you intellectually acknowledge God but lack any genuine passion, devotion, or sense of His presence. This apathy is a strategic victory for the enemy because it renders you an ineffective warrior. It is the spiritual equivalent of a C2 breakdown, where the soldier on the front line no longer receives or trusts the orders from headquarters. He achieves this by attacking your spiritual disciplines and fostering a sense of profound disconnect from the divine.

This C2 attack comes on two primary fronts:

1. Jamming the Frequency (The Noise of the World)
The first front is an electronic warfare campaign designed to drown out the still, small voice of God with an overwhelming amount of cultural and personal static. The enemy understands that a quiet, reflective heart is a receptive heart, so he works to eliminate quiet and reflection from our lives. As Jesus warned in the Parable of the Sower, the seed of the Word can be "choked by life's worries, riches and pleasures, and they do not mature" (Luke 8:14). The enemy's goal is to create as many of these "thorns" as possible.

Distraction through Busyness: He overloads our schedules with work, social commitments, and even good ministry activities. He convinces us that we are too busy for the one relationship that sustains all others. We become so focused on working *for* God that we have no time to be *with* God, a subtle but devastating form of mission sabotage.

Entertainment as a Weapon: He provides an endless buffet of addictive entertainment – social media, streaming services, video games, televised sports- that consume hours of our time and fill our minds with worldly chatter, making it difficult to hear God's frequency.

The strategic result of this constant noise is a degradation of our spiritual senses. We may still believe in God, but we lose our ability to clearly hear His guidance, feel His presence, and discern His will. We become spiritually deafened by the static.

2. Discrediting the Commander (The Lies of Suffering)

When jamming the frequency isn't enough, the enemy opens a second front: a psychological operation designed to attack the credibility of the Commander Himself. During a long and difficult siege – a season of suffering, chronic illness, or unanswered prayer – he whispers insidious lies designed to make you doubt God's goodness and His control. He uses your pain as "evidence" to support his propaganda:

"If God were truly good, He wouldn't allow this suffering."
"He is silent because He doesn't care."

This tactic is designed to make us misinterpret the very nature of our trials. The Apostle James provides the divine counterintelligence on this matter: "Consider it pure joy, my brothers and sisters, whenever you face trials of many kinds, because you know that the testing of your faith produces perseverance. Let perseverance finish its work so that you may be mature and complete, not lacking anything" (James 1:2-4). God's purpose in trials is to forge us into stronger, more complete warriors. The enemy's propaganda campaign is to convince us that these same trials are proof of God's abandonment. His ultimate goal is to achieve what Paul described in 2 Corinthians 4:4, where the "god of this age has blinded the minds of unbelievers," and in our case, to blind believers to the goodness and sovereignty of God in the midst of their pain. This leaves us truly orphaned on the battlefield, a state from which surrender is almost inevitable.

Counterintelligence:
Protecting the Connection

In a military siege, the single most important task for the defending force, beyond rationing ammunition and food, is to fight to keep the lines of communication open. A desperate message relayed to a relief force, a single piece of intelligence received from High Command, or an order that coordinates a breakout can mean the difference between survival and annihilation. Counterintelligence in this phase, therefore, is the active, disciplined, and often desperate battle to protect and strengthen your connection to God amidst the enemy's deliberate campaign to sever it.

The enemy's C2 attack is sophisticated. He will use the noise of the world to jam your spiritual frequencies and the pain of your circumstances to discredit the character of your Commander. Your counter-offensive must be equally strategic, focused on hardening your spiritual communications infrastructure and reinforcing your unwavering trust in God, regardless of the battlefield conditions.

Directive 1: Secure the Comms Channel
(Prioritizing Time with God)

This is the non-negotiable foundation of your defense during a siege. When you are under sustained pressure, your daily time with God is not a luxury or a religious duty; it is your most critical strategic appointment. It is your time to receive intelligence from the Word, transmit your requests and situation reports in prayer, and have your spirit recalibrated and reinforced by the presence of your Commander.

Make it a Priority, not a Possibility: You must treat this time with the seriousness of a strategic briefing. Schedule it. Guard it fiercely. If it means getting up earlier or turning off the television an hour earlier at night, that is the price of maintaining the connection. A soldier in a besieged fortress does not "forget" to listen for the radio; his life depends on it.

Declutter the Battlefield: Be ruthless in identifying and minimizing the sources of noise in your life. You are in a battle for silence. This may require turning off phone notifications, practicing digital sabbaths (a day or even an evening away from screens), or creating a designated "quiet zone" in your home where you can meet with God without distraction. You must actively fight for the quiet margin needed to hear His voice above the noise of the siege.

Directive 2: Verify the Commander's Character (Trusting in the Dark)

When the enemy uses your suffering as a psychological weapon to attack God's character, you must launch a counter assault based on what you *know* to be true, not what you *feel* to be true in the moment. Feelings are casualties of war; doctrine is the bedrock of your fortress.

Default to Doctrine: Your emotions are unreliable intelligence in a long siege. Your faith must be anchored to the unshakeable, unchanging doctrine of God's sovereignty, goodness, and love as revealed in the entirety of Scripture. When your circumstances scream that God has abandoned you, you must counter-broadcast with the truth of Romans 8:28, that God works all things for the good of those who love Him. When you feel unloved, you must assault that feeling with the truth of Romans 8:38-39, that nothing can separate you from His love.

Review Past Victories: Your personal history with God is a powerful intelligence file. Keep a journal or a record of God's past faithfulness. When you are tempted in the present darkness to believe He has forgotten you, you must review the written record of all the times He has come through for you before. This history of His faithfulness is a powerful weapon against the enemy's propaganda of the moment.

Directive 3: Practice Active Listening and Two-Way Communication

Communication is a two-way street. A siege is not the time to only be transmitting your own desperate pleas for help. It is the time to lean in and listen intently for the Commander's voice, however quiet it may seem.

Praying Scripture: When your own words fail and your emotions are raw, the most powerful prayers are often God's own words prayed back to Him. Praying the Psalms, for example, gives you a divine language for your lament, your anger, and your hope. This aligns your heart with His will and sensitizes you to His truth even when you feel nothing.

Listening Prayer: Intentionally spend time in silence. After you have poured out your heart to God, be still. Quiet your own mental chatter and ask the Holy Spirit to speak. Often, the most profound strategic guidance – a sense of peace, a new perspective, a specific verse brought to mind – comes in the quiet after the storm of our own requests. This is an advanced discipline, but it is critical for anyone under a prolonged siege. It is the act of a soldier who knows that sometimes the most important order is the one that is whispered, not shouted.

Biblical Assault: After-Action Reports on Maintaining the Connection

The stories of God's people are filled with sieges, and they provide clear case studies on the life-and-death importance of maintaining a connection with Command. These accounts reveal that in the midst of a long, wearying battle, the one who can continue to hear and obey the voice of their Commander survives, while the one whose communication lines are severed is doomed.

Case Study: Elijah at Horeb
(1 Kings 19) – Reestablishing the Comms Link

After his triumphant victory over the prophets of Baal on Mount Carmel, the prophet Elijah entered a sudden and severe siege. Jezebel issued a death warrant for him, and Elijah, the man who had just called down fire from heaven, was now fleeing for his life, gripped by fear and despair. The enemy's campaign shifted from a direct confrontation to a psychological one, designed to degrade his capabilities by severing his connection to his own mission and his God.

The C2 Breakdown: Elijah's report from the field was one of complete demoralization. He sat under a broom tree and prayed for death, saying, "I have had enough, LORD... I am no better than my ancestors" (1 Kings 19:4). He later complained to God, "I am the only one left, and now they are trying to kill me too" (1 Kings 19:10). His intelligence was faulty (there were 7,000 other faithful in Israel), and his spirit was broken. The enemy's psychological assault had succeeded in making him feel isolated and abandoned.

The Divine Counter Assault: God's response was not a booming command, but a tutorial in restoring a broken soldier. First, He attended to Elijah's physical needs with food and rest. Then, He brought him to Horeb, the mountain of God, to reestablish the comms link. God demonstrated that His voice was not in the overwhelming power of the wind, the earthquake, or the fire, but in a "gentle whisper" (1 Kings 19:12). This was the critical moment. In the midst of Elijah's emotional chaos, God bypassed the noise and spoke with quiet clarity.

The Strategic Result: Once the connection was reestablished, God gave Elijah three things: **New Intelligence** ("Yet I reserve seven thousand in Israel – all whose knees have not bowed down to Baal"), **a New Mission** (anoint two new kings and his own successor, Elisha), and a **Renewed Sense of Purpose.** God didn't rebuke Elijah for his fear; He reconnected with him, corrected his faulty intelligence, and redeployed him to the battlefield. The siege on Elijah's soul was broken when the line to his Commander was restored.

Case Study: King Saul's Final Battle (1 Samuel 28)

The tragic end of King Saul provides the ultimate example of a commander completely cut off from his C2. Facing a massive Philistine army at Shunem, Saul was terrified. He desperately sought guidance and a strategic plan, but his own actions had systematically degraded his capabilities.

The Severed Connection: The text is chillingly clear and serves as the final verdict on Saul's leadership: "He inquired of the LORD, but the LORD did not answer him – by dreams or Urim or prophets" (1 Samuel 28:6). Because of his persistent rebellion, his pride, and his direct disobedience over many years, Saul's line of communication to High Command was dead. He had been court-martialed by his own actions. He was an orphaned commander, left to face a superior enemy with no divine intelligence, no strategy, and no encouragement.

The Desperate Alternative: In his terror and desperation, having been completely isolated from his true Commander, Saul turned to a forbidden and corrupt source of intelligence: he disguised himself and sought out the witch of Endor to try and communicate with the dead prophet Samuel. This was a catastrophic act of spiritual treason. It is what happens when a commander on the battlefield can no longer reach headquarters and, in a panic, turns to the enemy's frequency, hoping for any signal at all.

The Strategic Consequence: The message he received only confirmed his doom. He went into his final battle not with hope, but with a prophecy of his own death. He and his sons were killed, and the army of Israel was routed. He was defeated because he had allowed his sin to completely degrade his capability to hear from God, proving that no amount of armor or soldiers can save a leader who has been cut off from his Commander.

Closing Charge: Keep the Line Open

You have seen the enemy's siege doctrine. His objective is to surround you, to wear you down, and to win by cutting you off from your Commander. He will jam your spiritual frequencies with the noise of the world. He will use the hardships of the siege to launch a propaganda campaign against the character of God Himself. He wants you to believe you are fighting alone. He wants you to think the silence you feel is abandonment.

But the line is never truly dead. Your Commander is always broadcasting. The challenge of the siege is to fight through the static, to reject the propaganda, and to strain to hear His voice. Your daily time in the Word and in prayer is not a religious duty; it is a vital communications check. It is the act of getting on the radio and confirming you are still connected to High Command.

Do not allow the enemy to degrade your capabilities. Do not allow the noise of the world or the pain of the battle to convince you to turn off your radio. Fight for the silence. Fight for the connection. In the grinding exhaustion of the siege, the single most important battle you will fight is the one to keep the line to your Commander open. Your survival, your endurance, and your ultimate victory depend on it.

Keep the line **open**.

Chapter 13

Destroying Key Infrastructure (The Foundations of Faith)

Military Warfare: The War on Foundations

IN A PROTRACTED war, a sophisticated adversary knows that defeating an army in the field is only one path to victory. A more insidious and often more permanent strategy is to bypass the army altogether and attack the nation's core infrastructure – the foundational systems that allow it to create, supply, and sustain its military forces. This is a strategic-level assault aimed at the enemy's heartland.[1] It targets power plants, transportation networks, communication hubs, and industrial centers. The goal is to cripple the enemy's ability to wage war by destroying the very foundations upon which that ability rests. An army without fuel, ammunition, or reinforcements is an army destined for collapse, no matter how courageous its soldiers.

Attacks on key infrastructure are designed to induce a state of national paralysis. Destroying a nation's power grid can shut down its factories and plunge its command centers into darkness. Severing its transportation networks – its bridges, railways, and ports – can prevent the movement of troops and supplies, isolating armies and leaving them to wither on the vine. Targeting its industrial base, such as oil refineries or munitions factories, directly degrades its ability to produce the essential materials of war. This is a deep, painful, and strategic form of warfare that aims not just to win the current battle, but to ensure the enemy cannot effectively fight any future battles either.

Case Study: The Allied Strategic Bombing Campaign (World War II)

Throughout the Second World War, the Allied forces, particularly the American Eighth Air Force and the British RAF Bomber Command, waged a relentless strategic bombing campaign against the industrial heart of Nazi Germany. While some operations were aimed at demoralizing the population, the core military objective was the systematic destruction of Germany's key war-making infrastructure.

This was a highly calculated campaign. Allied intelligence identified critical "choke points" in the German war economy. Planners knew, for example, that modern armies run on ball bearings; without them, you cannot build tanks, planes, or trucks. This led to a series of costly and famous raids on the ball bearing factories at Schweinfurt.[2] Similarly, the German military was powered by oil. The Allies therefore launched a massive campaign against the German oil industry, targeting synthetic fuel plants and oil refineries, most notably the massive complex at Ploesti in Romania.

These attacks were incredibly dangerous for the aircrews and did not produce an immediate victory. However, their cumulative effect was devastating. By relentlessly pounding Germany's infrastructure, the Allies severely hampered its ability to replace its losses.[3] The German Luftwaffe struggled to build new fighter planes, the Panzer divisions were often short on fuel, and the entire war machine was slowly ground down. The strategic bombing campaign was a direct assault on the foundations of the Third Reich's military power, contributing significantly to its eventual collapse.

Case Study: Sherman's March to the Sea (U.S. Civil War, 1864)

In 1864, the Union General William Tecumseh Sherman executed a campaign to destroy the enemy's logistical and psychological infrastructure. After capturing the vital city of Atlanta, Sherman did not turn to fight the main Confederate army under General Hood. Instead, he embarked on his infamous "March to the Sea," cutting a 60-mile-wide path of destruction across Georgia to the port of Savannah.

Sherman's objective was explicit: to destroy the Confederacy's ability and will to continue the war.[4] His forces were ordered to live off the land and to destroy all assets that could support the Confederate war effort. They systematically tore up railroad lines, heating the rails and twisting them around trees to create "Sherman's neckties."[5] They burned farms, seized food supplies, and destroyed factories. This was not a battle against an army; it was a war on the infrastructure that sustained that army.

The psychological impact was just as important as the physical destruction. The march demonstrated to the people of the Confederacy that their government could not protect them. It showed that a Union army could march with impunity through the heart of their territory, severing the link between the soldiers at the front and their homes. It was a devastating blow to Southern morale and is widely considered to be a key factor in precipitating the end of the Civil War. Sherman didn't just defeat an army; he broke the back of the society that supported it.

Spiritual Warfare: Attacking the Foundations

In the midst of a spiritual siege, when frontal assaults on our will have failed, Satan often adopts the more insidious strategy of attacking our spiritual infrastructure. He knows that if he cannot force a quick surrender, he can achieve the same result by systematically dismantling the very foundations upon which our faith is built. He attempts to create spiritual "blackouts" by attacking our core beliefs, to isolate us from our resources by attacking our spiritual disciplines, and to disrupt the pathways of grace by attacking our trust in God. He doesn't need to destroy the soldier if he can first destroy the soldier's belief in the cause, his trust in his Commander, and the supply lines that sustain him.

This is a subtle but devastating form of siege warfare. The enemy doesn't always come with an obvious, banner-waving temptation. He comes as a philosopher, a critic, a "realist," sowing seeds of doubt designed to erode our core convictions over time. He lays siege to the temple of our faith (1 Corinthians 3:17) by launching a systematic, strategic bombing campaign against its four foundational pillars:

The Foundation of Truth (Our Belief in Scripture)

The first target is always our source of intelligence and doctrine: the Bible. The enemy whispers questions about its trustworthiness, its historical accuracy, its relevance to modern life, and its divine authority. He promotes the idea that it is an outdated book, biased by its human authors, or open to endless subjective interpretation. He seeks to damage our confidence in this primary source of truth, so that when we are in a crisis, we no longer trust the only map we have. If he can degrade our belief in the Word of God, he can cut us off from our most essential supply of truth, leaving us to navigate the fog of war with nothing but our own flawed feelings and the enemy's deceptive whispers.

The Power Source (Our Conviction in the Gospel)

Next, he attacks the very engine of our spiritual lives: the Gospel itself. This is a two-front assault. On one front, he subtly twists the Gospel to make it more palatable to the world, watering down its essential truths. He promotes a "Gospel-plus" narrative – that you need grace *plus* good works, or grace *plus* a particular spiritual experience, or grace *plus* adherence to a political ideology. On the other front, he bombards believers, especially those struggling with sin, with feelings of intense unworthiness. He works to make them feel that the atoning sacrifice of Christ is not sufficient for *their* specific failures, that they have abused grace one too many times. By corrupting our understanding of the

Gospel, he seeks to sever our connection to the very power that saved us and sustains us. This violates the foundation of grace we see in Ephesians 2:8-9 that it is by grace *alone* that we are saved.

The High Command (Our Trust in God's Goodness)

This is the enemy's primary psychological weapon during a extended siege. He uses the painful reality of our circumstances – protracted suffering, unanswered prayer, tragic loss – to make us question God's goodness, His love, and His sovereign control. He relentlessly whispered lies rooted in our pain:

o   *"If God really loved you, He wouldn't let this happen."*
o   *"If God were all-powerful, He would have changed this by now."*
o   *"His silence is proof that He has abandoned you."*

His goal is to create a sense of deep disillusionment and spiritual despair, causing us to doubt the very character of our Commander. It is critical in these moments to discern between an attack and a test. As the Apostle James states, God uses trials to produce perseverance and make us "mature and complete" (James 1:2-4). The enemy's strategy is to take a divine test designed to strengthen our faith and twist it into an attack designed to demolish it.

The Supply Lines
(Our Commitment to Spiritual Disciplines)

Finally, the enemy launches a war of attrition against our spiritual supply lines: our dedication to prayer, Bible reading, fellowship, and service. He knows these practices are essential for our spiritual nourishment and our connection with God. During a long siege, he works to make these disciplines seem burdensome, irrelevant, or ineffective. He whispers:

o   *"What's the point of praying? Nothing is changing."*
o   *"You're too tired to read your Bible; it's not doing any good anyway."*
o   *"You should skip church this week; you need a break from all those people."*

This is the spiritual equivalent of Sherman's March to the Sea – he seeks to cut us off from our sources of spiritual sustenance, leaving us weak, starved, and vulnerable in the face of his continued assault.

Counterintelligence: Fortifying the Foundations

To withstand a siege designed to destroy your infrastructure, you must become an expert in spiritual civil engineering. Meaning, your counterintelligence mission is to actively inspect, maintain, and fortify the foundational pillars of your faith. You cannot take them for granted during peacetime and hope they will stand during a war. As the military axiom goes, "He who sweats in training, bleeds less in battle." You must treat your core beliefs and disciplines as the strategic assets they are, guarding them with the vigilance and intentionality of a commander protecting his most critical command bunkers, power plants, and supply depots.

The enemy's assault on your infrastructure is subtle and patient. He will not blow up the bridge; he will slowly corrode the steel until it collapses under pressure. He will not launch a frontal assault on the fortress; he will poison the well. Therefore, your defense must be equally patient, proactive, and deeply rooted.

Directive 1: Reinforce Your Doctrinal Foundation (Belief in Scripture)

Your trust in the Word of God is the bedrock upon which your entire spiritual fortress is built. If this foundation cracks, everything else will eventually crumble. You must be proactive in defending and reinforcing your confidence in Scripture.

Study Its Reliability: Do not be afraid to engage with the intellectual side of your faith. A soldier who understands the mechanics of his rifle is more confident in its use. Learn about the historical evidence for the Bible's accuracy. Understand the incredible manuscript evidence that supports its transmission through the ages. Familiarize yourself with basic apologetics so that when the enemy whispers doubts about the Bible's trustworthiness, you have a well-reasoned, logical, and faith-filled answer.

Live by Its Authority: The ultimate proof of the Bible's power is its effect on a transformed life. The more you intentionally submit to the Word, obey its commands, and see its fruit in your own life, the more your confidence in its divine origin will grow. Living out the truth is the most powerful way to fortify your belief in it.

Directive 2: Anchor Yourself in the Gospel (Protecting Your Power Source)

The Gospel is not just the entry gate to the Christian life; it is the nuclear reactor that powers the entire city. The enemy will constantly try to dilute its power or convince you that you are unworthy to access it.

Meditate on Grace Daily: You must preach the Gospel to yourself every single day. Start your mornings by reminding yourself that your standing with God is based entirely on the *finished* work of Christ, not on your performance from the day before. This protects your heart from the twin lies of pride (when you do well) and despair (when you fail).

Reject "Gospel-Plus" Doctrines: Be vigilant against any teaching that adds to the work of Christ. The enemy loves to introduce the idea that you need "grace *plus* " something else – good works, a special spiritual experience, adherence to a political ideology. The simple, unadorned Gospel is your power source; guard it fiercely against any who would try to tamper with it.

Directive 3: Defend God's Character (Countering Enemy Propaganda)

When you are in a long and painful siege, your feelings will lie to you. The enemy will use those feelings as a vehicle for his propaganda, slandering the character of your Commander. Your countermove is to trust God's revealed character over your present circumstances.

Know His Attributes: You must study the attributes of God – His sovereignty, His omniscience, His goodness, His faithfulness, His justice. When your situation seems chaotic and out of control, you must anchor your mind to the doctrinal truth that He is sovereign. When He seems distant and uncaring, you must hold fast to the revealed truth that He is good and that nothing can separate you from His love.

Document His Past Faithfulness: Your personal history with God is a powerful intelligence file. Keep a journal of answered prayers, moments of divine provision, and past deliverances. In times of intense doubt and suffering, reviewing this written, historical record of God's good character in your own life is a powerful act of defiance against the enemy's lies.

Directive 4: Maintain Your Spiritual Disciplines (Guarding the Supply Lines)

In a siege, spiritual disciplines are not optional duties; they are critical lifelines. The enemy wants you to neglect them because a starved soldier cannot fight. You must fight to maintain them, especially when you feel like it the *least*.

Fight for Prayer: Even when it feels like you are talking to the ceiling, maintain the discipline of prayer. This is you keeping the radio on, refusing to let the enemy cut your line to Command.

Feed on the Word: Even when it feels dry and you feel you get nothing out of it, maintain the discipline of reading Scripture. This is your daily ration. It is the spiritual sustenance that is keeping you alive for the long fight, whether you feel its effects immediately or not.

Don't Abandon Fellowship: The enemy wants to use your struggle to isolate you. Resist this with everything you have. Be honest with your trusted allies about the attacks on your foundations. Let them support you, pray for you, and speak truth to you. Your fellowship is the reinforcement convoy that can break a siege.

Biblical Assault: After-Action Reports on Foundational Warfare

The enemy's strategy of attacking the foundational infrastructure of faith is a well-documented tactic. These case studies show how Satan attempts to demolish the core beliefs of God's people and how a resolute defense of that foundation is critical for survival and victory.

Case Study: The Attack on Job's Theology (Book of Job)

The story of Job is the ultimate siege. After the initial, devastating assaults on his family, wealth, and health, Satan's main attack shifted to a prolonged siege against his spiritual infrastructure. The primary weapons in this siege were the "friends" of Job, who launched a relentless assault on his core beliefs.

The Attack on Infrastructure: Job's friends, operating from a flawed and simplistic theology, systematically attacked his trust in God's justice and goodness. Their argument was a constant refrain: "You are suffering, therefore you must have sinned." This was a direct, sustained attack on the very foundations of Job's faith. They were trying to destroy his understanding of God, replacing it with a transactional, karma-like system where all suffering is direct punishment for a specific sin. They assaulted his conviction in his own integrity before God, trying to force a false confession that would have validated their corrupt theology.

The Biblical Counter Assault: Job's defense was messy but resolute. He refused to have his infrastructure destroyed. He fiercely defended his own integrity, but more importantly, he refused to accept their flawed, slanderous view of God. Though he wrestled, questioned, and lamented, he never abandoned his core trust in God's ultimate sovereignty. His famous cry, **"I know that my Redeemer lives, and that in the end he will stand on the earth" (Job 19:25),** was an act of profound counter assault. It was a declaration that even with all his physical and emotional infrastructure in ruins, his foundational hope in a redeeming God would not be demolished. He held onto the foundation, even when the rest of the building was crumbling around him.

Case Study: The Corinthian Church and the Gospel (1 Corinthians 15)

The Apostle Paul discovered that a hostile force had infiltrated the church at Corinth and was launching a direct assault on the most critical piece of spiritual infrastructure: the doctrine of the resurrection. Some in the church had begun to teach that there was no resurrection of the dead, attacking the very "power plant" of the Christian faith.

The Attack on Infrastructure: This was not a minor theological disagreement. Paul understood that if you destroy the foundation of the resurrection, the entire structure of the Gospel collapses. He lays out the catastrophic consequences of this infrastructural attack: if Christ is not raised, "our preaching is useless and so is your faith," "we are found to be false witnesses about God," "you are still in your sins," and "we are of all people most to be pitied" (1 Corinthians 15:14-19). Satan, through these false teachers, was attempting a strategic demolition of their central belief.

The Biblical Counter Assault: Paul's response in 1 Corinthians 15 is a prime example of reinforcing spiritual infrastructure. He did not just rebuke them; he systematically re-laid the foundation with logic, evidence, and divine truth.

He Restated the Core Doctrine: He began by reminding them of the simple, historical Gospel he first delivered: "that Christ died for our sins according to the Scriptures, that he was buried, that he was raised on the third day..." (v. 3-4).

He Provided Eyewitness Intelligence: He listed the verifiable witnesses to the resurrected Christ – Peter, the Twelve, more than 500 brothers, James, all the apostles, and finally himself. This was the intelligence report, the undeniable proof.

He Explained the Strategic Importance: He then explained the catastrophic consequences of removing this foundational piece, showing them how the entire system of their faith would collapse without it.

Paul's assault was to counter an attack on the foundation by methodically and powerfully rebuilding that foundation with logic, eyewitness testimony, and passionate theological reasoning, ensuring the fortress of their faith would stand.

Closing Charge: Defend the Foundation

You have seen the enemy's doctrine for a long and grinding siege. When he cannot defeat you with a quick assault, Satan will lay siege to your soul and begin a systematic campaign to destroy your spiritual infrastructure. He will attack the authority of the Word you stand on. He will attack the truth of the Gospel that saved you. He will use your pain to attack the goodness of the God you serve. He will use your weariness to attack the disciplines that sustain you. He is not just fighting the soldier; he is trying to demolish the fortress from its foundations up.

But these foundations have been laid by God Himself, and they are not so easily destroyed. Your charge, in the midst of the siege, is to become a vigilant guardian of this holy infrastructure.

o   When he attacks the Word, double down on your study and stand on its promises.

o   When he attacks the Gospel, preach its grace to yourself every morning.

o   When he attacks God's character in your suffering, wage a counter-offensive of worship and recall His past faithfulness.

o   When he attacks your spiritual disciplines, fight for that time of prayer and Scripture reading as if it were your last ration of food – because it is.

Do not be the foolish man who builds his house on the sand, only to see it collapse in the storm. Be the wise man who builds his house on the rock. The winds of the siege will blow, the rains of suffering will fall, and the flood of the enemy's lies will rise against you. If your foundations are neglected, you will fall. But if you have been diligent in guarding and reinforcing them, you will stand.

Defend the foundation.

Chapter 14

Attrition Warfare (Wearing Us Down)

Military Warfare: The Strategy of the Grind

WHEN A SWIFT, decisive offensive fails and a lengthy war begins, the nature of the conflict often shifts to one of history's most grueling and brutal strategies: attrition warfare. This is the grim mathematics of the battlefield, a tactic that seeks victory not through brilliant maneuver or tactical surprise, but through the systematic and deliberate process of grinding down an enemy's manpower, material, and will to fight until they can no longer offer effective resistance. It is a war of numbers, a test of endurance, and a contest of resolve. The commander who employs an attrition strategy is betting that he can absorb and inflict losses at a rate that is unsustainable for his adversary.[1]

Attrition warfare is not characterized by elegant flanking maneuvers, but by bloody, head-on confrontations and relentless, sustained pressure. It is the Somme and Verdun in World War I, where hundreds of thousands of men were thrown into a meat grinder for a few yards of muddy ground. It is the Battle of Stalingrad, where entire armies were consumed in house-to-house fighting.[2] The objective is not necessarily to seize a key piece of terrain, but to destroy the enemy's army in place. It is a haunting calculation that accepts massive casualties on one's own side as a necessary price for inflicting even greater, and ultimately unbearable, casualties on the other.

This strategy targets every aspect of an enemy's ability to wage war. It is a constant barrage of artillery to induce shell shock and exhaustion. It is a continuous series of small-scale attacks and raids designed to keep the enemy constantly on alert, denying them rest or the ability to reorganize. It is a relentless assault on supply lines, ensuring that every bullet, every meal, and every gallon of fuel that reaches the front is paid for in blood. The psychological goal of attrition is to create a persistent sense of hopelessness, to convince the enemy soldier that the fighting will never end, that his replacements will also be consumed, and that his cause is ultimately doomed.

Case Study: The Battle of the Atlantic (World War II)

The longest continuous military campaign of the Second World War was a classic and devastating example of attrition. From the very first day of the war, Nazi Germany sought to defeat Great Britain by starving it into submission. As an island nation, Britain was completely dependent on its maritime supply lines – the convoys that brought food, oil, and war materiel from North America and the rest of its empire. The German strategy, executed by their U-boat fleet, was to sink these merchant ships faster than the Allies could replace them.[3]

This was a brutal war of statistics. The "tonnage war," as it was called, was a grim ledger of ships sunk versus ships built. German U-boats, operating in "wolf packs," stalked the Atlantic, inflicting horrific losses on Allied convoys. The Allies responded with improved convoy tactics, new anti-submarine technologies like sonar and radar, and a massive shipbuilding program. For years, the outcome was in doubt. Winston Churchill famously stated that the "only thing that ever really frightened me during the war was the U-boat peril."[4] He knew that Britain could withstand the Blitz, but it could not survive without its maritime lifeline. The Allies ultimately won the Battle of the Atlantic not in a single, decisive naval engagement, but by grinding down the U-boat fleet, sinking submarines faster than Germany could build them, and by producing ships at a rate that eventually outstripped the losses.[5] It was a victory of attrition, a triumph of industrial capacity, and endurance over a lethal and determined foe.

Case Study: The Vietnam War (U.S. Strategy)

The American military strategy in the Vietnam War was, for much of the conflict, a deliberate strategy of attrition. Unable to launch a full-scale invasion of North Vietnam due to political constraints, the U.S. commander, General William Westmoreland, pursued a "search and destroy" policy. The objective was not to capture and hold territory in the traditional sense, but to inflict such heavy casualties on the Viet Cong and the North Vietnamese Army (NVA) that they would lose their will and ability to continue the fight.

This led to a war where the primary metric of success was the "body count."[6] U.S. forces would use their superior mobility and firepower to engage enemy units, kill as many as possible, and then withdraw. The belief was that there was a "crossover point" at which the U.S. was killing enemy soldiers faster than the North could replace them.[7] However, this strategy was deeply flawed. It underestimated the enemy's incredible resolve and willingness to absorb staggering losses for their cause. It also failed to account for the political and psychological nature of the war. While the U.S. consistently won the tactical battles, the nightly images of American casualties on television eroded support for the war back home. Ultimately, the North Vietnamese and Viet Cong won the war of attrition not by outfighting the Americans, but by outlasting them. They proved that in a war of attrition, the side with the greater *will* to endure can defeat a foe with far greater firepower.

Spiritual Warfare:
Attrition Warfare (Wearing Us Down)

This is a war of exhaustion, a grinding, relentless campaign designed not for a swift, dramatic victory, but for a slow, suffocating defeat. Satan understands that a soldier who can withstand a single, mighty blow may yet crumble under the weight of a thousand relentless cuts. The objective of his attrition warfare is to methodically deplete our spiritual, emotional, and even physical reserves until we become too weary to fight. He doesn't need to force a catastrophic moral failure; he simply needs to make us so tired that we **stop resisting, stop caring, and stop actively pursuing God.**

This strategy is a masterful exploitation of fleshly weakness. While the believer's spirit is made new in Christ, we still operate within the confines of a limited body and mind. The enemy's attrition campaign is characterized by a constant, often undetectable, bombardment of pressures that target our resolve, our joy, and our focus. He seeks to create a state of perpetual low-grade crisis, a spiritual static that drains our energy and erodes our will to persevere.

The Sustained Bombardment of Temptation

In the Initial Assault phase, temptation often arrives like a sudden artillery strike – overt, powerful, and designed to shock us into submission. In attrition warfare, however, temptation becomes a constant, rolling barrage of small-arms fire. The enemy's goal is to *normalize* sin by turning the entire cultural landscape into a hostile environment. As the "ruler of the kingdom of the air," he works tirelessly to shape the world's systems to be conducive to his goals (Ephesians 2:2). He strengthens temptation by making sin readily accessible, intellectually defensible, and culturally celebrated. He glorifies "the lust of the flesh, the lust of the eyes, and the pride of life" (1 John 2:16) through media that portrays immorality as empowerment, materialism as success, and rebellion as enlightenment.

This creates an atmosphere where our spiritual defenses are under constant assault. Every billboard, every television show, every social media feed can become a delivery system for temptation. The enemy's strategy is to create so much noise that the signal of sin becomes background noise, something we grow accustomed to and eventually stop guarding against. He knows that while God is faithful to provide a way out of every temptation (1 Corinthians 10:13), a soldier who is fighting on a thousand fronts at once is more likely to miss the escape route on one of them. This is not about a single moment of weakness, but about the cumulative effect of a thousand moments of exposure, designed to wear down our moral vigilance and make compromise seem like a reasonable, even necessary, reprieve from the constant fight.

Psychological Operations: The War on the Mind

Simultaneously, Satan launches a sophisticated psychological warfare (PSYOP) campaign directly against our minds. This is the whispering campaign of doubt, fear, and accusation that plays on a loop in the back of our thoughts. He is a master of exploiting our individual vulnerabilities, which he meticulously studied during his pre-conflict reconnaissance. He attacks our minds with a relentless stream of negative thoughts, anxieties, and lies, seeking to destabilize our emotions and cloud our judgment. His objective is to demolish our confidence not in ourselves, but in God. He plants insidious seeds of doubt:

- *"If God really loved you, He wouldn't let this happen."*
- *"You've failed again; you'll never be free from this."*
- *"Your prayers are hitting the ceiling."*
- *"God is disappointed in you."*

These are not random thoughts; they are targeted, weaponized lies designed to cut our trust in God's goodness and power. As Paul commands, we are to "take captive every thought to make it obedient to Christ" (2 Corinthians 10:5), precisely because the enemy's primary theater of operations is the mind. He seeks to create a fortress of fear, anxiety, and self-condemnation within us, an internal enemy that saps our strength and leaves us feeling isolated, hopeless, and emotionally exhausted. An emotionally depleted soldier cannot effectively wield their weapons.

The Campaign of Discouragement and Disillusionment

Attrition warfare's most effective weapon may be discouragement. Satan exploits the gap between our expectations and our reality. He magnifies our failures, highlights the slow pace of our spiritual growth, and points to the apparent success of the ungodly as "proof" that our faith is futile. He wants to take our journey of sanctification – a lifelong process – and frame it as a series of failures. He will constantly remind you of past sins, even those forgiven by God, to keep you weighed down by shame and guilt.

This is where he "cuts in" on our race, seeking to keep us from obeying the truth (Galatians 5:7). He takes the trials that God intends for our refinement and twists them into evidence of God's absence. He exploits delayed answers to prayer, persistent struggles, and the general weariness of life in a fallen world to build a case for disillusionment. The goal is to extinguish the fire of our "first love," to replace passionate devotion with tired obligation, and eventually to convince us to settle for a spiritual ceasefire – a comfortable but lukewarm faith that poses no threat to his kingdom. He wants us to believe that the abundant life Jesus promised (John 10:10) is unattainable, so we stop fighting for it.

Exploiting Physical and Logistical Weakness

Finally, the enemy's attrition strategy recognizes the undeniable link between our physical and spiritual states. He knows that a tired body leads to a vulnerable spirit. He pushes us toward a lifestyle of overwork, busyness, and neglect of our physical health. He doesn't need to cause the illness or the exhaustion; he is an expert at exploiting the spiritual openings they create. When we are physically drained, our resolve weakens, our minds are less clear, and we are far more susceptible to temptation and discouragement. Neglecting sleep, proper nutrition, and rest is akin to a soldier voluntarily cutting their own supply lines. The enemy rejoices when we are too busy to pray, too tired to read the Word, and too stressed to connect with fellow believers. This physical depletion is a force multiplier for all his other attrition tactics, making his bombardment of temptation and his psychological operations exponentially more effective.

In summary, Satan's attrition warfare is a multifaceted, patient, and grinding assault on our endurance. He layers temptation upon discouragement, and anxiety upon exhaustion, hoping that the cumulative weight will eventually cause us to buckle. The goal is not a dramatic takedown but a slow fade into combat ineffectiveness. Recognizing this insidious, long-game strategy is the critical first step toward developing the counter-strategies necessary to not only survive the siege, but to outlast the enemy and stand firm until the final victory is declared.

Counterintelligence: Cultivating Spiritual Stamina

Countering a war of attrition is not about a single, heroic counter-offensive; it is about cultivating the discipline and resilience to outlast the enemy. A soldier can be equipped with the finest weapons, but if their supply lines are cut, their morale collapses, and their body fails from exhaustion, they will be defeated without the enemy firing a single decisive shot. Therefore, our counterintelligence strategy against Satan's attrition warfare must be focused on proactive replenishment, disciplined rest, and the intentional fortification of our spiritual, mental, and physical endurance. To withstand this grinding siege, we must become masters of spiritual logistics.

The following directives are not optional exercises for when you feel strong; they are non-negotiable survival protocols for the long war. They are the means by which we "do not become weary in doing good," so that "at the proper time we will reap a harvest if we do not give up" (Galatians 6:9).

Directive 1: Conduct Regular Readiness Assessments (Know Your Limits)

A foolish soldier ignores their dwindling ammunition and pushes forward until their weapon is empty. A wise soldier constantly assesses their state and knows when they need to fall back to a secure position to reload. You must learn to pay meticulous attention to your internal dashboard. The enemy's attrition campaign is designed to be subtle, and you must become skilled at recognizing its effects before they become critical.

Acknowledge Your State: Be ruthlessly honest with yourself about your physical, emotional, and spiritual condition. Learn to recognize the warning signs of weariness: increased irritability, difficulty concentrating, feelings of overwhelm, a pervasive sense of hopelessness, or a loss of joy in things that once brought you life. These are not signs of failure; they are the blinking red lights on your control panel, signaling an urgent need to rest and recharge.

Reject the "Tough It Out" Mentality: Our culture, and sometimes even our church culture, can glorify burnout as a sign of devotion. This is a lie the enemy exploits. Ignoring your limits is not a sign of strength; it is a strategic vulnerability. Acknowledging your weariness is the first step toward addressing it and preventing the enemy from using your exhaustion as a weapon against you.

Directive 2: Maintain and Secure Your Supply Lines (Proactive Replenishment)

In a siege, the flow of supplies is everything. Satan's attrition seeks to cut you off from your spiritual and physical resources. Your counter strategy must be to fiercely guard and consistently access your supply lines.

Spiritual Fuel (The Word and Prayer): Your daily time in prayer and Bible study are not religious duties; they are the logistical convoys that deliver your spiritual fuel and rations. You must make them non-negotiable priorities. Prayer for endurance, wisdom, and a renewed sense of God's presence is like calling in air support to relieve pressure on your position. Reading the Word is like receiving your daily bread; it provides the spiritual nutrients you need to keep going, even when you don't feel an immediate effect. A soldier who stops eating will eventually be too weak to lift his rifle. A Christian who stops taking in the Word will become too weak to resist.

Physical Readiness (Strategic Rest): Your body is not a machine; it is the physical vessel in which you fight this spiritual war. The enemy knows this and will exploit your physical exhaustion. You must view rest, proper nutrition, and sleep not as luxuries, but as critical components of your combat readiness. Taking time to rest and recharge is not a retreat from the battle; it is a *strategic withdrawal* to a fortified position for resupply and reinforcement. A well-rested soldier is more alert, more resilient, and far less susceptible to the enemy's psychological operations.

Directive 3: Defend Against Psychological Warfare (Fortify Morale)

Attrition warfare is fundamentally a war on morale. Satan wants to use discouragement, shame, and disillusionment to convince you to lay down your arms. You must counter this PSYOP campaign with the unshakable truth of the Gospel.

Remember God's Grace: Satan will constantly replay the highlight reel of your failures. You must counter this broadcast by deliberately meditating on the truth of God's grace. Your past mistakes and current struggles do not define you. Cling to the promise that His grace is always available and that you are not, and never will be, too far gone for His restoration (1 John 1:9).

Break the Power of Shame and Guilt: When you sin or fall short, the enemy wants you to wallow in shame and self-condemnation. Your standing order is to confess it to God immediately. This act breaks the enemy's power. It rejects the feelings of shame and guilt he is trying to use to control you and instead activates the cleansing power of Christ's blood.

Replace Self-Condemnation with Self-Compassion:
Treat yourself with the same grace you would offer a fellow soldier who is wounded and struggling. Recognize that you are human, that the battle is real, and that you will face setbacks. This is not an excuse for sin, but an acceptance of the reality of sanctification in a fallen world.

Directive 4: Practice Active Gratitude
(Acknowledge Victories and Reinforce Hope)

In a long, grinding campaign, it is easy to focus only on the mud, the losses, and the exhaustion. Gratitude is a powerful counter-weapon that forces you to shift your focus.

Celebrate Small Wins: Actively look for and celebrate small victories. Did you resist a temptation today? Did you choose kindness when you felt angry? Did you spend five minutes in prayer when you felt too tired? Acknowledge these as victories. Celebrating these wins builds momentum and reminds you that you are making progress.

Document God's Faithfulness: Keep a running log of God's goodness in your life. Write down answered prayers, moments of unexpected joy, and instances of His provision. This becomes your personal testimony, a written record that you can review during times of intense weariness. It is a powerful reminder that your Commander is faithful and that your struggle is not in vain. There is a reason this is mentioned more than once – it works.

Ultimately, the goal of withstanding attrition is found in Paul's command: "Therefore put on the full armor of God, so that **when** the day of evil comes, you may be able to stand your ground, and after you have done everything, to stand" (Ephesians 6:13). By cultivating spiritual stamina through these directives, you will be equipped not just to survive the enemy's war of exhaustion, but to remain standing, weapon in hand, ready for the next engagement.

Biblical Assault: After-Action Reports on the War of Endurance

The strategy of attrition is one of Satan's most frequently deployed and effective weapons. He knows that if he cannot achieve a swift victory through a frontal assault, he may still win by simply outlasting our will to fight. Scripture provides critical case studies of this grinding warfare, not only showing us the enemy's tactics in high definition but also revealing the divine counter strategies for enduring and overcoming a campaign of pure exhaustion. These are not just stories of hardship; they are blueprints for survival in a long war.

Case Study: Elijah Under the Juniper Tree (1 Kings 19) – The Attrition of Despair

No soldier is more vulnerable to attrition than one who is depleted after a major victory. The story of Elijah following his triumph on Mount Carmel is a textbook example of a post-engagement spiritual collapse. Elijah had just executed a flawless spiritual assault, single-handedly confronting 450 prophets of Baal, proving the power of God before the entire nation, and seeing fire fall from heaven. He was at the absolute peak of his operational effectiveness.

The Attrition Attack: Satan did not need to mount a new offensive. He simply needed to exploit the massive physical, emotional, and spiritual energy Elijah had just expended. The counter attack came not from an army, but in the form of a single, credible threat from Queen Jezebel: "May the gods deal with me, be it ever so severely, if by this time tomorrow I do not make your life like that of one of them" (1 Kings 19:2). After facing down an entire false religion, this one threat was enough to break him. The enemy's timing was perfect. Elijah, utterly spent, collapsed. The Bible says he "was afraid and ran for his life." He ran a day's journey into the wilderness, sat down under a broom tree, and prayed for death: "I have had enough, LORD... Take my life; I am no better than my ancestors" (1 Kings 19:4). This is the ultimate goal of attrition warfare: to bring a soldier to the point of total surrender, to make them believe the fight is no longer worth fighting.

The Divine Counter Strategy: God's response to Elijah's complete burnout is a masterclass in countering attrition. He did not give him a lecture or a new battle plan. He initiated a program of strategic logistical support. An angel appeared and provided for Elijah's most basic needs: sleep and nourishment. The angel woke him, gave him bread and water, and let him sleep again. Then a second time, the angel came, saying, "Get up and eat, for the journey is too much for you" (1 Kings 19:7). God acknowledged his servant's profound weariness. Only after his physical strength was restored did God address his spiritual and emotional despair, not with a rebuke, but with a gentle question: "What are you doing here, Elijah?" (1 Kings 19:9). God allowed him to vent his hopelessness and countered the lie of isolation by revealing, "Yet I reserve seven thousand in Israel – all whose knees have not bowed down to Baal" (1 Kings 19:18). The counter assault was not a command to "be tougher," but a demonstration of divine care: rest, provision, presence, and a renewed sense of purpose.

Case Study: Nehemiah and the Wall
(Nehemiah 4 & 6) – The Attrition of Combined Arms

When Nehemiah led the project to rebuild the walls of Jerusalem, he faced a relentless, multi-pronged attrition campaign designed to halt the work by exhausting the workers.

The Attrition Attack: The enemies, Sanballat and Tobiah, understood they could not defeat the Jews in a direct, righteous confrontation, so they resorted to a siege of morale.

Psychological Warfare (Mockery): They began with a barrage of ridicule designed to demoralize the builders. "What are those feeble Jews doing?... Can they bring the stones back to life from those heaps of rubble – burned as they are?" (Nehemiah 4:2). The goal was to make the work seem futile.

Threat of Violence: When mockery failed, they escalated to threats, plotting to attack Jerusalem (Nehemiah 4:8). This forced the builders into a state of constant alert, draining their energy. They had to work with one hand and hold a weapon in the other, a physically and mentally exhausting posture.

Internal Fatigue: The enemy's strategy was amplified by the very nature of the task. The people of Judah themselves reported, "The strength of the laborers is giving out, and there is so much rubble that we cannot rebuild the wall" (Nehemiah 4:10). The enemy was using the difficulty of the mission itself as a weapon.

Sustained Distraction: Finally, they tried to wear down the leader. Four separate times, they sent a message to Nehemiah, trying to lure him away for a "meeting" (Nehemiah 6:2-4). This was a campaign of distraction and harassment, hoping to break his focus and exhaust his resolve.

Nehemiah's Counter Strategy: Nehemiah's leadership provides the field manual for withstanding this kind of assault.

Immediate Prayer: His first response to every threat was to turn to his Commander: "Hear us, our God, for we are despised" (Nehemiah 4:4).

Practical Defense: He combined prayer with action, posting guards and arming the workers. He did not wait for a divine miracle while ignoring practical defense.

Rallying Morale: He actively countered the psychological warfare by reminding the people of their higher purpose and their powerful Commander: "Don't be afraid of them. Remember the Lord, who is great and awesome, and fight for your families, your sons and your daughters, your wives and your homes" (Nehemiah 4:14).

Unyielding Focus: He decisively rejected every attempt at distraction, sending back the same unwavering message: "I am carrying on a great project and cannot go down" (Nehemiah 6:3). He refused to let the enemy dictate the terms of engagement.

These after-action reports teach us a vital lesson: withstanding attrition requires both divine dependence and disciplined determination. Like Elijah, we must be willing to receive God's physical and spiritual provision. And like Nehemiah, we must relentlessly focus on our mission, refuse to be demoralized by mockery or distracted by harassment, and actively fortify the morale of those around us. This is how we fulfill the command to "not become weary in doing good," so that we may "reap a harvest if we do not give up" (Galatians 6:9).

Closing Charge: Outlast the Enemy

You have been shown the enemy's grinding doctrine of attrition. This is not the lightning strike of a shocking temptation, nor the overt assault on your core beliefs. This is the long, weary siege – a war of exhaustion designed with a single, brutal objective: **to make you quit**. Satan knows that a fortress that cannot be taken by storm can still be starved into submission. He will use the relentless pressure of trials, the constant hum of temptation, and the heavy weight of discouragement to drain your resolve, deplete your resources, and extinguish your will to fight.

His campaign is one of slow erosion. He wants you to believe the lie that your small struggles are a sign of total failure. He wants you to mistake weariness for weakness. He wants to convince you that the battle is unwinnable, that your Commander is distant, and that surrender is the only path to peace. He is counting on your fatigue to become your philosophy.

But you have been given the counterintelligence. You know that victory in a war of attrition is not won by the soldier who is strongest on the first day, but by the one who is still standing on the last. This is where your spiritual disciplines cease to be duties and become your very survival. Your time in the Word is not a checklist item; it is your supply line. Your prayer is not a formality;

it is your call for reinforcement. Your rest is not a luxury; it is a strategic necessity to refit and rearm.

Therefore, do not despise the moments when, like Elijah, you are utterly spent. Your Commander sees your exhaustion and provides for your recovery. But do not yield to the hopelessness that exhaustion brings. Instead, let the stubborn resilience of Nehemiah become your model. When the enemy mocks, when threats fly, when your own spirit grows tired, you are to pray, work, and keep one hand on your sword.

The ultimate command in this phase of the war is found in Ephesians 6:13: "and after you have done everything, to stand." Your charge is not to feel strong. Your charge is not to feel victorious. Your charge is simply to endure. Outlast the enemy's campaign of lies. Outlast his barrage of discouragement. Outlast his assault on your hope. In the grinding war of attrition, endurance is its own form of attack.

Do not grow weary.

Stand **firm**.

Shane Cunningham

Chapter 15

Securing Key Terrain
(Claiming Territory in Our Lives)

Military Warfare: The Battle for a Foothold

WAR, IN ITS most elemental form, is a violent argument over geography.[1] Armies do not fight in a vacuum; they fight on, for, and over the ground itself. Within this unforgiving calculus, not all ground is created equal. There are patches of earth – hills, bridges, cities, crossroads, and mountain passes – that hold a value far exceeding their physical size. This is "key terrain": ground that, if seized and held, confers a decisive, disproportionate advantage to its possessor. The battle for key terrain is the battle for control, the struggle for a foothold from which the war can be won.[2]

A commander who understands strategy does not see a map of hills and rivers; they see a landscape of opportunities and threats. A hilltop is not just a mound of dirt; it is an observation post that can see for miles, a dominant firing position that can rain down destruction with impunity. A bridge is not just a structure of steel and concrete; it is the throat through which an army's supplies and reinforcements must flow. A city is not just a collection of buildings; it is a hub of transportation, a center of communication, and a potent symbol of national will. To control this terrain is to control the movement, the vision, and often the morale of the entire battlefield.[3]

Case Study: The Battle of Hamburger Hill (1969)
In May 1969, during the Vietnam War, U.S. forces launched a major assault on Hill 937 in the A Shau Valley – a heavily fortified North Vietnamese Army (NVA) stronghold. Though the hill itself held little intrinsic value, it commanded the surrounding valley and offered critical observation and artillery advantages. For ten days, American paratroopers of the 101st Airborne Division fought uphill through dense jungle, monsoon rains, and entrenched defenses. Casualties were severe, earning the battle its grim nickname: "Hamburger Hill," because soldiers felt they were being ground up like meat in the fight.[4]

Despite immense losses, the hill was finally taken on May 20th. Within weeks, however, U.S. forces withdrew, leading many to question whether the cost had been worth the prize.[5] Yet the struggle underscored the brutal truth about key terrain: commanders will sacrifice dearly to seize and hold ground that can dominate the battlefield. Even a single hill, if it provides an enemy with a stronghold and line of sight, can shape the course of an entire campaign.[6]

Spiritual Warfare: The Battle for a Foothold

Just as an army fights for hills and bridges, our adversary fights for territory within our very souls. The battle for spiritual "key terrain" is the battle for our hearts, minds, and wills. Satan's ultimate objective during the siege is not merely to harass us from the outside, but to establish a fortified, operational base *inside* our defenses. He seeks to move beyond temptation and into occupation. He knows that a soldier can fight an external enemy for years, but an enemy that has established a fortress behind the lines can cause chaos, disrupt supply lines, and demoralize the soldier into submission from within.

This campaign begins with a "foothold." The Apostle Paul issues a direct command regarding this tactic: "and do not give the devil a foothold" (Ephesians 4:27). A foothold is a small, seemingly insignificant piece of ground ceded to the enemy through unaddressed sin, unforgiveness, agreement with a lie, or trauma. It is an unlocked door, a gap in the wall. By itself, it may seem minor. But to a strategic enemy, a foothold is not an end; it is a beachhead. It is the beginning of an invasion. If this ground is not immediately reclaimed, the enemy will not give it back. He will fortify it. He will dig in, build walls, and erect a fortress. This fortified territory, built on the ground of an uncontested foothold, is what Scripture calls a **stronghold.**

Satan doesn't just want to tempt us to sin; he wants to establish these strongholds in our lives – fortified areas where sin, negative beliefs, and destructive patterns have taken root and gained control. As the Apostle Paul writes, we are not to "offer any part of yourself to sin as an instrument of wickedness" (Romans 6:13). A stronghold is precisely that: a part of our being – our mind, our emotions, our habits – that has been turned into a weaponized instrument for the enemy's use. These strongholds act as his forward operating bases. From these secure positions, he can influence our thoughts, manipulate our emotions, and dictate our behaviors, making it incredibly difficult to live a life that pleases God.

These enemy fortifications are not built of stone and mortar, but of arguments, lies, and deceptions. The intelligence on this is explicit: "We demolish arguments and every pretension that sets itself up against the knowledge of God" (2 Corinthians 10:4-5). Strongholds are constructed from these very things:

Arguments: These are the sophisticated rationalizations for sin we build in our minds. *"This isn't really hurting anyone." "I deserve this." "It's just how I'm wired."* These justifications act as the defensive walls of the stronghold, repelling the truth of God's Word.

Pretensions (Lofty Opinions): These are the towers of pride and false beliefs that elevate our own ideas or feelings above the knowledge of God. This can be a deeply rooted belief like, *"I'm worthless and unlovable,"* which stands in direct opposition to the truth of our identity in Christ. It can also be a theological pretension like, *"God's grace covers this, so I don't need to change,"* which twists truth into a license for sin.

These strongholds manifest in our lives as the key terrain the enemy has successfully seized:

Fortresses of Recurring Sin and Addiction: This is terrain where a specific sin is no longer just a temptation, but a ruling power. It could be lust, substance abuse, rage, or gossip. The believer feels trapped in a cycle of defeat, repeatedly falling into the same pattern despite their best efforts. This is because they are not just fighting a temptation; they are launching an assault against a fortified enemy position that exists *within them*.

Garrisons of Negative Beliefs: These are mental strongholds where the enemy has successfully garrisoned his lies. Crippling fear, pervasive anxiety, deep-seated bitterness, or a victim mentality can become the lens through which we see the world. A believer living in a stronghold of fear is constantly under the influence of that enemy garrison, making decisions and reacting to life from a place of torment, not truth.

Bunkers of Destructive Patterns: This is territory where our emotional and relational habits have been co-opted. This could be patterns of self-sabotage, codependency in relationships, or an inability to trust. These ingrained patterns function like enemy bunkers, providing cover from which he can fire on our relationships, our purpose, and our peace.

When Satan secures this key terrain, the believer is no longer fighting for freedom from an external position. They are living under a partial and voluntary occupation. Christ has set us free, but by allowing a stronghold to remain, we willingly place ourselves "back under a yoke of slavery" (Galatians 5:1). This is the enemy's goal in the siege: to take a free soldier of Christ and, by securing territory within their life, turn them into an effective prisoner of war, neutralized from within and unable to fight for the Kingdom. Recognizing the existence of these enemy fortresses on our own soil is the first, critical step in preparing the counter assault to demolish them and reclaim every inch of our lives for the glory of God.

Counterintelligence: Demolishing Enemy Fortifications

When an enemy has successfully seized key terrain *inside* your perimeter, the defensive posture changes. This is no longer a battle fought at the walls; it is a counter offensive to retake ground that has been lost. To allow an enemy fortress to stand unchallenged within your own territory is strategic suicide. It must be identified, isolated, and systematically demolished. The counterintelligence for this phase of the war is not passive; it is an active, aggressive, and divinely empowered demolition campaign.

The Apostle Paul provides our mandate and our hope: "The weapons we fight with are not the weapons of the world. On the contrary, they have divine power to demolish strongholds" (2 Corinthians 10:4). This is your operational guarantee. You are not fighting with willpower alone; you have been issued spiritual munitions powerful enough to level any fortress the enemy has built in your life. This process is deliberate, courageous, and requires a series of coordinated actions.

Directive 1: Conduct Internal Reconnaissance (Identify the Stronghold)

Before you can demolish a fortress, you must have its exact coordinates. You must become a ruthless intelligence officer of your own soul. This requires brutal honesty and a willingness to look at the patterns the enemy has successfully established. You cannot defeat what you will not define.

Map the Territory: Prayerfully ask God to reveal the key terrain the enemy has seized. Ask the hard questions: Where do I feel most powerless? What is the recurring sin that traps me in a cycle of defeat and shame? What negative belief about myself, God, or the world do I constantly return to? What is the source of the persistent fear, anxiety, or anger in my life?

Identify the Triggers: What situations, people, or thought patterns consistently lead you into the stronghold's sphere of influence? Recognizing these triggers is like identifying the supply routes to the enemy's fortress.

Trace Its Origins: How was this stronghold built? Was it through an unconfessed sin that gave the enemy a "foothold"? Was it through believing a lie spoken over you years ago? Was it built on the foundation of trauma? Understanding its construction helps you know how to dismantle it.

Directive 2: The Two-Part Assault
(Confession and Renunciation)

Once the stronghold is identified, the assault begins. This is a two-pronged attack that breaks the enemy's legal right to the territory and reestablishes God's rightful authority.

Confession (Realigning with God): The first act is to confess any sin associated with the stronghold directly to God. As 1 John 1:9 promises, "If we confess our sins, he is faithful and just and will forgive us our sins and purify us from all unrighteousness." Confession is not about groveling; it is a strategic act of agreeing with God about the sin, stepping out of the darkness of the stronghold and into the light of His truth. It realigns you with your Commander.

Renunciation (Evicting the Enemy): Confession deals with your sin; renunciation deals with the enemy. This is a verbal, authoritative act. You must specifically name the stronghold (e.g., "the stronghold of fear," "the stronghold of lust," "the stronghold of bitterness") and declare that you are breaking your agreement with it. You renounce its influence and command it to go in the name of Jesus Christ. You are serving an eviction notice, declaring that this territory is no longer under enemy control but belongs to the Kingdom of God.

Directive 3: Call for Reinforcements
(Enlist Allies and Air Support)

No soldier is expected to single-handedly assault a fortified enemy position. It is foolish and dangerous. You must break the enemy's primary siege tactic of isolation.

Enlist Your Allies: Find trusted, mature believers – a pastor, a mentor, a small group – your Battle Buddy – and bring your struggle into the light. Vulnerability is a weapon. Asking for help and accountability brings in ground support, reinforcing your position and ensuring the enemy cannot operate in secrecy.

Request Air Support: This is fervent, targeted prayer for deliverance. Ask God to send His power to break the stronghold's grip. Pray for healing from the wounds that allowed it to be built. This is not just your prayer; when you enlist allies, their prayers join yours, creating a concentrated bombardment of the enemy's position.

Directive 4: Rebuild and Refortify
(Renewing the Mind)

Demolishing a stronghold leaves a void. If you do not deliberately rebuild on that reclaimed ground, the enemy will attempt to reoccupy the "empty house" (Matthew 12:44-45). This is the critical reconstruction phase.

Counter with Truth: Identify the specific lies that formed the stronghold's foundation and counter them with the specific truths of Scripture. If the stronghold was built on the lie "I am worthless," you rebuild with the truth "I am fearfully and wonderfully made" (Psalm 139:14) and "I am God's masterpiece" (Ephesians 2:10). Write these truths down. Memorize them. Meditate on them.

Occupy Your Mind: You must actively "take captive every thought to make it obedient to Christ" (2 Corinthians 10:5). This means consciously choosing to fill your mind with what is true, noble, right, pure, lovely, and admirable (Philippians 4:8). This is an act of spiritual occupation, filling the reclaimed territory with the presence of God's kingdom.

Directive 5: Maintain Vigilance (Continuous Security)

Once a strategic hill is taken, a wise commander immediately sets up a permanent guard post. You must do the same. Be aware that the enemy will test your new defenses. He will try to reestablish his influence. Standing firm, as commanded in Galatians 5:1, is an ongoing posture. By practicing these new patterns of thought and behavior, you fortify the ground you have reclaimed, transforming what was once an enemy fortress into a bastion of God's strength in your life.

Consider the second day of the Battle of Gettysburg in 1863. The Union Army held a defensive line along a series of ridges. On the extreme left flank was a small, rocky hill known as Little Round Top. It was, for a time, unoccupied. From this key terrain, Confederate artillery could have enfiladed the entire Union line, shattering their defenses and potentially winning the battle, and with it, the war. When a Union general spotted the impending disaster, the desperate race to seize and hold this hill began. The ferocious defense of Little Round Top by the 20th Maine, culminating in a dramatic bayonet charge, is now a hallowed moment in military history precisely because everyone involved understood the stakes.[7] They were not fighting for a pile of rocks; they were fighting for the piece of ground that controlled the destiny of the nation.

The process of securing key terrain is a three-stage operation. First is **identification.** This is an intelligence-driven process of analyzing the battlefield to determine which locations are critical. It requires a commander to think ahead, to see not just where the enemy is, but where they *need to be* to win. Second is **seizure.** This is often the most violent and costly phase. It requires a focused, overwhelming assault – an airborne drop behind enemy lines to capture a bridge, an amphibious landing to establish a beachhead, or a concentrated armored thrust to take a vital crossroads. It is a high-risk, high-reward gambit to gain a foothold in enemy territory.

Finally, and most critically, is **consolidation and holding.** Once key terrain is taken, the battle is not over; it has just begun. The capturing force must immediately fortify their new position, digging trenches, setting up fields of fire, and preparing for the inevitable, furious counter attack. The enemy, knowing the strategic value of the ground they have just lost, will throw everything they have into taking it back. The ability to hold this newly acquired territory under extreme pressure is what transforms a temporary gain into a permanent strategic advantage. It becomes a secure base of operations, a forward operating base from which all future offensive actions can be launched, allowing an army to project power deep into what was once the enemy's heartland. The struggle for key terrain is, therefore, the essential battle for a foothold – the bloody, necessary work of claiming and holding the ground from which victory becomes possible.

Biblical Assault: After-Action Reports on Demolition Campaigns

The divine power to demolish strongholds is not a theoretical concept; it is a demonstrated reality woven throughout the pages of Scripture. These case studies serve as our field manual for liberation, providing detailed debriefings on how fortified enemy positions – both in the lives of individuals and in the path of God's people – have been decisively leveled by the power of God. These are not just miracles; they are tactical precedents for the demolition campaigns we are called to wage today.

Case Study: The Gerasene Demoniac
(Mark 5) – Liberating Occupied Territory

Building on our discussion of Satan's hierarchy in chapter 4, we now analyze the "Legion" account from the perspective of occupied territory. Within the New Testament, no other narrative provides such a stark and detailed portrayal of an enemy stronghold. This was far more than a struggle with fleeting temptation – it was the complete subjugation of a human being, transforming him into fully occupied enemy territory.

The Enemy Stronghold: The man's life was a fortress of demonic power. He lived among the tombs, a place of death, completely cut off from community – a classic siege tactic of isolation. He was beyond the control of any human authority, breaking chains and shackles. He was engaged in constant self-harm, a sign of the enemy's core mission to "steal and kill and destroy" (John 10:10). The name of the occupying force, "Legion," was a chillingly accurate military term, implying a force of thousands of demonic entities garrisoned within a single soul. This man was the very definition of key terrain seized and held by the enemy.

The Divine Assault: Jesus Christ, our Commander-in-Chief, arrives on the scene and launches a direct, authoritative assault.

Direct Confrontation: Jesus did not wait for an invitation. He took the offensive, approaching the stronghold head-on.

Authoritative Command: His initial weapon was a command: "Come out of this man, you impure spirit!" (Mark 5:8). He immediately asserts His superior authority.

Forced Reconnaissance: Jesus then demands intelligence: "What is your name?" (Mark 5:9). This forces the enemy to identify itself, exposing the scale of the occupation and stripping away its anonymity.

Decisive Eviction: Jesus does not negotiate or compromise. He grants their request to enter the pigs only to demonstrate their ultimate powerlessness and destructive nature before Him. He issues the final command, and the occupying legion is evicted. The territory is liberated.

The Aftermath: Rebuilding and Redeployment: The result was a total reversal of the occupation. The townspeople find the man "sitting there, dressed and in his right mind" (Mark 5:15). What was once a screaming, violent fortress of chaos was now a peaceful defender of God's restorative power. The ground was not only reclaimed but immediately refortified with sanity and peace. The man was then given a new mission: to become a witness to the power of God in the region, turning a former enemy base into a forward operating post for the Gospel.

Case Study: The Walls of Jericho (Joshua 6) – Demolishing an External Fortress

Before Israel could possess the Promised Land, they had to deal with the fortress city of Jericho. This was the primary piece of key terrain that blocked their path. It was a physical stronghold, heavily fortified and humanly impenetrable.

The Enemy Stronghold: Jericho's walls represented a massive, intimidating barrier. The city was "securely shut up," a symbol of the enemy's power to deny God's people their inheritance. From a conventional military perspective, a direct assault would have been costly and likely futile for a nomadic people without advanced siege weaponry.

The Divine Assault: God's battle plan was a direct assault on human logic and military pride. It was a spiritual operation from start to finish, demonstrating that "the weapons we fight with are not the weapons of the world" (2 Corinthians 10:4). The strategy involved:

Obedient Repetition: Marching around the city once a day for six days. This was an act of disciplined, patient faith.

Focus on God's Presence: The Ark of the Covenant held the central position in the procession, a tangible symbol of God's very presence. This was not merely a ceremonial display; it underscored the profound truth that victory would be achieved through God's power, not through human strength or military prowess. The battle's outcome rested entirely on divine intervention, a testament to the unwavering belief in God's omnipotence.

The Final Assault: As the seventh day arrived, the instructions became even more specific: seven circuits followed by a long, resonant trumpet blast. Then, the people were to shout. This was no ordinary battle cry; this was a spiritual act, a unified shout of faith intended to unleash God's power against the stronghold. The battle wasn't physical, but spiritual, and their shout was the key to unlocking divine intervention.

The Aftermath: Total Demolition: The result of this unconventional, faith-based assault was catastrophic for the enemy. "The wall collapsed; so everyone charged straight in, and they took the city" (Joshua 6:20). God demonstrated that any fortress, no matter how imposing, will crumble when confronted with the obedient, faith-filled actions of His people wielding the spiritual weapons He provides.

These after-action reports confirm our operational doctrine. Whether the stronghold is an internal fortress of sin and torment or an external barrier of opposition, our weapons – the authority of Christ's name, the power of His presence, and the active exercise of our faith – have been divinely designed to bring them crashing down.

Closing Charge: Reclaim the Territory

You have seen the enemy's ultimate objective in the siege: not merely to defeat you, but to *occupy* you. He seeks to move beyond harassment and into colonization, to build his fortresses on your holy ground, and to raise his flag over territory that Christ has already purchased with His blood. A stronghold of bitterness, a garrison of fear, a fortress of addiction – these are not minor tactical problems; they are acts of illegal occupation that cannot be tolerated.

This is a violation of the terms of your liberation. A soldier freed from a POW camp does not allow the enemy to set up a guard post in his living room. You were set free by Christ; you are not to tolerate a single inch of enemy occupation. The very ground on which these strongholds stand belongs to your Commander, and He has given you the authority and the weapons to reclaim it.

You have been issued divine munitions, as stated in 2 Corinthians 10:4, with the power to level any fortress the enemy has built. The stories of Jericho and Legion are not just history; they are your operational precedents. They are proof that no wall is too high and no occupying force is too great to be evicted by the power of God.

Therefore, the mission is clear. Conduct your reconnaissance. Identify the enemy's positions within your heart and mind. Launch the assault with the weapons of confession, renunciation, and the power of God's Word. Call in your allies for support. Reclaim the territory. Tear down the flags of the enemy and raise the banner of Christ over every area of your life. There is no peaceful coexistence with an occupying force.

Demolish the strongholds.

Chapter 16

Isolating the Enemy
(Isolating Us from Christ)

Military Warfare:
The Anatomy of Annihilation

IN THE BRUTALITY of war, few situations are more dreaded or terrifying than the radio call that confirms a unit is cut off and surrounded. This is the military doctrine of isolation, of encirclement – a maneuver designed not merely to defeat an enemy, but to annihilate it. It is the deliberate act of severing a fighting force from the lifeblood of its parent army, transforming it from a lethal, integrated component into a doomed pocket of resistance.[1] An army that can successfully isolate a portion of the enemy force has already, in most cases, sealed its fate. The ensuing battle is often just a grim formality.

The strategy of encirclement is as ancient as organized conflict but was perfected in the 20th century with the German concept of the *Kesselschlacht (KES-əl-shlaht),* or "cauldron battle."[2] The operational art is both elegant and savage. Swift, armored spearheads punch through the enemy's front line at two separate points, driving deep into their rear. These pincer movements then pivot towards each other, linking up to close the ring and trap entire divisions, corps, or even whole armies inside the newly formed "kessel" *(KES-əl)*, or cauldron. Once this ring of steel is closed, the surrounded force is no longer fighting a war of objectives; it is fighting a desperate, hopeless war of pure survival.

An isolated unit is a dying unit, because it is systematically starved of the three elements essential for its existence:

Logistics (The Severing of Supply): A modern army is a voracious consumer of resources. It requires a constant, massive flow of ammunition, fuel, food, medical supplies, and spare parts. When the supply lines are severed, this flow stops instantly. The effect is not immediate but is as certain as gravity. Every bullet fired, every artillery shell launched, every gallon of fuel burned is one that can never be replaced. The army's logistical "tail" has been cut, and its combat "tooth" begins to weaken. The force begins a slow, inexorable starvation.

Reinforcements (The Denial of Manpower): In the friction of combat, casualties are a certainty. A healthy army absorbs these losses and replenishes its ranks with fresh troops from the rear. An encircled army has no rear. Every soldier killed or wounded is a permanent deletion from the combat roster. The unit begins to bleed out, its fighting strength diminishing with every passing hour and every skirmish fought. The line becomes thinner, the defenders more exhausted, and the inevitable collapse draws nearer.

Command and Control (The Collapse of Morale): Perhaps most devastating is the psychological impact. While a surrounded unit may maintain radio contact for a time, its meaningful connection to the larger strategic picture is gone. It can no longer coordinate its actions with a relief force. Orders from high command become increasingly irrelevant to the terrifying reality on the ground. Hope for rescue dwindles, replaced by a crushing sense of abandonment. Morale – the spiritual glue that holds an army together – disintegrates under the immense psychological pressure of being utterly alone and marked for death.

At the Battle of Cannae in 216 B.C., the Carthaginian general Hannibal, though vastly outnumbered, executed a brilliant double envelopment of the Roman army, encircling and annihilating a force of over 50,000 men in a single afternoon.[3] It remains a textbook masterpiece of tactical encirclement. Millennia later, at Stalingrad in the winter of 1942, the Soviet Operation Uranus achieved the same result on a monstrous scale, trapping the entire German Sixth Army – nearly 300,000 soldiers – in a frozen cauldron.[4] Cut off from all ground supply, subjected to the brutal Russian winter, and relentlessly attacked from all sides, the once-invincible army was methodically starved, frozen, and crushed into submission over three agonizing months.

The strategic imperatives of this doctrine are absolute. For an attacking force, the encirclement of a significant enemy unit is a supreme operational prize, a shortcut to victory. For the defending force, the cardinal rule is to never allow its units to be cut off. To be isolated on the battlefield is to be handed a death sentence. It may be carried out slowly through starvation and attrition, or quickly in a final, bloody assault, but the outcome is rarely in doubt. The goal of this tactic is to turn a functioning, integrated part of a larger army into a helpless island of despair, ripe for total destruction.

Spiritual Warfare: Cutting the Unit Off from the Army

The enemy's doctrine of encirclement has a terrifying and precise parallel in the spiritual realm. Satan, our adversary, understands the military principle of isolation better than any human general. He knows that while a single believer is a soldier, a community of believers is an army. Therefore, his most strategic, high-value objective during a protracted siege is to sever the individual soldier from their unit – the Body of Christ. He seeks to surround, cut off, and create a "kessel," or cauldron, of isolation around a believer, knowing that once they are cut off from their spiritual supply lines, reinforcements, and command structure, their annihilation is not a matter of *if*, but *when*.

An isolated believer is a dying believer. Their spiritual "logistics" – the life-giving supply of encouragement, wisdom, and corporate worship that flows from the community – are severed. Their "reinforcements" – the prayers of their brothers and sisters and the accountability that keeps them strong – can no longer reach them. Their connection to "command and control" – the guidance and correction from pastors and mature saints – is jammed. The ultimate goal of this strategy is to take a vibrant, integrated member of Christ's army and turn them into a helpless, starving island of despair. As the wisdom of Scripture states, "Pity anyone who falls and has no one to help them up!" (Ecclesiastes 4:10). The enemy's entire strategy is to engineer a situation where you are that person.

To achieve this encirclement, Satan deploys a multifaceted and insidious campaign designed to break the bonds of fellowship. His signature move, as always, is deception (John 8:44), and here he uses it to turn us against our most vital allies.

The Offensive of Offense (Creating the Breach): The enemy's primary tool for initiating isolation is offense. He knows the church is not a museum of perfect saints, but a hospital for the broken. It is filled with imperfect people who will inevitably say and do things that hurt one another. The enemy exploits these moments, taking a genuine hurt or a minor slight and nurturing it into a full-blown grievance. He whispers justifications for anger and elevates our personal pain into a righteous cause for withdrawal. This is how the "bitter root" that Hebrews warns about takes hold, a root that, if left unaddressed, "grows up to cause trouble and defile many" (Hebrews 12:15). This bitterness becomes the enemy's initial breach in the defensive perimeter of fellowship, a "foothold" (Ephesians 4:27) that justifies pulling away from the very community that could bring healing.

The Propaganda of Prideful Individualism (The "Lone Wolf" Lie): Satan has launched a masterful psychological operation against the modern church, rebranding isolation as a sign of spiritual strength. He whispers lies that appeal directly to our pride:

- *"You don't need them."*
- *"The church is full of hypocrites; you're better off on your own."*
- *"Your relationship with God is personal; you don't need organized religion."*

This is a lie packaged in the appealing wrapper of self-sufficiency. He convinces the believer that they are a spiritual "lone wolf," strong and independent. In reality, a lone wolf separated from the pack is not a predator; it is prey. This lie directly contradicts the scriptural design for the church as a "body," where each part is essential and codependent (1 Corinthians 12:27).

Fueling Internal Division (Turning the Camp Against Itself): Where envy and selfish ambition exist, Scripture warns, "there you find disorder and every evil practice" (James 3:16). The enemy works tirelessly to sow these seeds within the church. He incites arguments over non-essential doctrines, promotes gossip that erodes trust, and fosters a critical spirit that sees flaws in every leader and fellow member. He creates factions, cliques, and an "us versus them" mentality *within the same unit*. This internal strife fractures the army from the inside, making it impossible for believers to "carry each other's burdens" (Galatians 6:2) because they are too busy fighting one another. A divided church cannot stand, and its soldiers become easy targets for the enemy to pick off one by one.

Weaponizing Shame and Inadequacy (Exploiting Past Defeats): For the believer who is worn down by attrition (Chapter 14) or is struggling with an internal stronghold (Chapter 15), the enemy deploys shame as his final, deadly encirclement tool. He whispers:

- *"If they truly knew what you struggle with, they would reject you."*

o *"You are too messy, too sinful, too broken for this perfect community."*

He takes the very struggles that should be brought into the light for healing and uses them as a reason to retreat into the darkness of isolation. This lie is diabolical because it drives the wounded soldier away from the hospital, convincing them that their wounds make them unworthy of care.

The enemy's strategic objective is clear: by using offense, pride, division, and shame, he aims to methodically cut you off from the Body of Christ. He wants to sever you from the flow of life, strength, and support that God designed for your survival. To be cut off from the Body is to be cut off from your life support. Recognizing this overarching strategy is the first and most critical step in resisting it and fighting for the fellowship that is not just a blessing, but a battlefield necessity.

Counterintelligence:
Breaking the Encirclement

When a military unit finds itself surrounded, its operational priority shifts entirely. Every action becomes focused on a single, desperate objective: breaking the encirclement. This is not a passive defense; it is an aggressive, all-out effort to reestablish contact with the main army and reopen supply lines. To accept isolation is to accept annihilation. Our counterintelligence against the enemy's spiritual siege mirrors this reality. Fighting for fellowship is not a social preference; it is a tactical imperative of the highest order.

Satan's entire strategy of isolation hinges on convincing you that the fight for community is too costly, too painful, or simply

unnecessary. He wants you to believe the lie that you are better off alone. Our counter offensive, therefore, must be a deliberate, disciplined, and relentless refusal to be cut off. We must treat our connection to the Body of Christ with the same ferocity a soldier would fight to protect their last radio or their final supply route.

Directive 1: Fortify Your Position in the Main Army (Commit to the Local Church)

A soldier without a unit is a wanderer, not a warrior. Your primary fighting formation is the local church. The enemy's "lone wolf" propaganda is designed to make you see this commitment as optional. You must see it as your lifeline.

Reject the Consumer Mindset: Do not treat church as a product to be consumed, but as a covenant to be honored. You are not there to be entertained; you are there to be integrated into a fighting force. Show up. Serve. Give. Participate even when it's inconvenient. Your consistent presence strengthens the entire unit and reinforces your own position against the enemy's attempts to pick you off.

Submit to the Command Structure: A healthy church has God-ordained leadership. Respecting and praying for your pastors and elders is a direct assault on the enemy's spirit of rebellion and division. It reinforces the divine command structure that brings order and strength to the army of God.

Directive 2: Maintain Internal Cohesion (Practice Radical Forgiveness and Grace)

An army that is fighting itself is an army the enemy has already defeated. The enemy's primary weapon for creating an encirclement is offense. Therefore, forgiveness and grace are your primary tools for breaking out.

Make Forgiveness Your Standard Operating Procedure: You will be hurt by people in the church – it is a guarantee. When offense comes, you must recognize it as an enemy attack designed to justify your withdrawal. Your standing order is to "be willing to forgive others who have wronged you" and "not let bitterness and resentment fester." Forgiveness is not a feeling; it is a tactical decision to disarm the enemy's chosen weapon. It neutralizes the bitter root before it can grow and defile the unit (Hebrews 12:15).

Extend Grace as a Defensive Shield: The church is a field hospital, not a parade ground for perfect soldiers. You must "extend grace and understanding to others, recognizing that we are all on a journey of growth." A critical, judgmental spirit is a corrosive agent that eats away at unity. Choosing to see your brothers and sisters through the lens of grace protects your own heart from pride and protects the fellowship from the enemy's attempts to fracture it.

Directive 3: Open a Secure Communications Channel (Embrace Vulnerability)

The enemy's siege thrives in darkness and secrecy. He uses shame over your struggles and strongholds to convince you to hide, thus completing his encirclement. The most powerful way to break this siege is to do the one thing he is banking on you never doing: speak.

Resist the Urge to Hide: Your struggles and wounds do not disqualify you from the fellowship; they are the very reason you need it. You must "resist the urge to hide" and instead "seek help when needed." Talk to a pastor, a mentor, or your Battle Buddy. Vulnerability is a courageous act of counter attack. It brings the enemy's secret operations into the light and allows the power of the community to be brought to bear on your situation. As Galatians 6:2 commands, we must "carry each other's burdens," and a burden cannot be carried if it is hidden.

Directive 4: Provide Covering Fire and Mutual Support (Fight for Your Unit)

In a firefight, soldiers provide covering fire for one another, protecting their comrades while they move or reload. In the spiritual battle, our prayers and support are the covering fire that protects our unit.

Pray for Other Believers: Actively "make a habit of praying for other believers." Pray for their protection, their strength, and their victory over the enemy. This transforms your relationship with them from a passive acquaintance into an active, mutual defense pact.

Be the Reinforcement: The promise of Ecclesiastes 4:10 is that when one falls, "one can help the other up." Look for the soldier who is struggling, who is isolated, or who is under fire, and be their reinforcement. An encouraging word, a helping hand, a listening ear – these are the logistical supplies that can keep a fellow soldier in the fight.

To be a Christian is to be a member of the Body of Christ (1 Corinthians 12:27). This is not a suggestion; it is a statement of fact. To fight for this unity is to fight for your own survival and for the victory of the entire army.

Biblical Assault: After-Action Reports on Coalition Warfare

The principle of coalition warfare is not merely a human strategy; it is a divine one, woven into the fabric of God's redemptive plan. Scripture is filled with accounts of how God's people, when united in a holy coalition, achieved victory, and when isolated or divided, faced catastrophic defeat. These are not just stories of community; they are tactical debriefings on the non-negotiable power of a unified front.

Case Study: The Jerusalem Command (Acts 2 & 4) – The Power of a Unified Front

In the aftermath of Christ's ascension, the early church was a small, vulnerable unit deep in hostile territory. They faced the religious establishment that had just crucified their leader and the immense power of the Roman Empire. From a purely military perspective, they were a prime target for encirclement and annihilation. The enemy's logical strategy would have been to scatter them through fear and intimidation.

The Biblical Counter Assault: Instead of scattering, the early believers executed a perfect counter strategy of consolidation. They did not just tolerate one another; they launched an aggressive assault *on isolation itself.* Their operational doctrine is laid out in Acts 2:42: "They devoted themselves to the apostles' teaching and to fellowship, to the breaking of bread and to prayer." This was not a casual social club; this was the disciplined life of a cohesive combat unit. They shared resources ("selling their property and possessions to give to anyone who had need"), maintained constant contact ("they continued to meet together in the temple courts"), and established secure internal supply lines ("they broke bread in their homes and ate together").

The Result: Unleashing Overwhelming Force: This radical unity became a force multiplier that unleashed devastating spiritual power. When the enemy counter-attacked with threats, the believers did not retreat into individual prayer closets. They gathered *together* as a single unit and launched a unified appeal to their Commander (Acts 4:24). The result was a spiritual earthquake: "the place where they were meeting was shaken. And they were all filled with the Holy Spirit and spoke the word of God boldly" (Acts 4:31). Their coalition warfare directly resulted in a fresh wave of divine power and combat effectiveness. The after-action report is clear: "With great power the apostles continued to testify" (Acts 4:33). Their unity was not just defensive; it was the engine of their offense.

Case Study: The Corinthian Infiltration
(1 Corinthians 1-3) – Catastrophe of a Broken Coalition

The church at Corinth serves as a grim warning – a case study in what happens when the enemy's strategy of isolation and division succeeds. This was a church blessed with spiritual gifts but was rendered almost completely combat-ineffective because Satan had successfully infiltrated their ranks and fractured their unity.

The Enemy's Assault: The enemy did not need to launch a massive external attack on the Corinthian church, because they were destroying themselves from within. He exploited their pride and "selfish ambition," leading to the very "disorder and every evil practice" that James 3:16 warns about. Paul's intelligence report is stark: "There are quarrels among you... One of you says, 'I follow Paul'; another, 'I follow Apollos'; another, 'I follow Cephas'; still another, 'I follow Christ'" (1 Corinthians 1:11-12). This was a unit that had turned its weapons on itself. They were a coalition in name only.

The Result: Combat Ineffectiveness: The consequences were devastating. A divided army cannot fight. Their internal squabbles led to spiritual immaturity, moral compromise (1 Corinthians 5), and an arrogant pride that blinded them to their own weakness. They were isolated from each other by their factions, and as a result, they were ineffective in their mission. Paul had to spend his time and energy correcting their internal failures rather than directing them in an outward assault on darkness.

These reports present a clear and absolute tactical principle. The unified, devoted, and interdependent community seen in Acts is a spiritual fortress against which the enemy's assaults break. The fractured, prideful, and inwardly focused community in Corinth is a broken unit, ripe for defeat. To fight for the unity of the Body is to fight for the victory of the army.

Closing Charge: Break the Encirclement

You have been shown the enemy's most cherished battlefield doctrine: isolation. He is a master of the *Kesselschlacht* – the cauldron battle. His goal is to encircle you, to sever your connection to the main army of God, and to leave you to starve, alone, in a pocket of despair. He will use the poison of offense, the propaganda of prideful individualism, and the shame of your personal struggles as his pincer movements to cut you off.

He wants to convince you that your brothers and sisters in Christ are the source of your frustration, not the source of your strength. He wants you to believe you are a strong, independent "lone wolf." He knows the truth: a wolf separated from the pack is not a predator; it is prey. He knows that once your supply lines of encouragement, accountability, and corporate prayer are severed, your defeat is only a matter of time.

Therefore, your charge is to declare war on isolation. You must see fellowship not as a social option, but as a tactical imperative for your survival. You must view the local church not as a building you visit, but as the fighting unit to which you belong. Let the dysfunctional, defeated church at Corinth be your warning. Let the world-shaking power of the unified church in Acts be your objective.

The command from headquarters is not merely to attend church; it is to *fight for your church*. Fight against the spirit of division. Fight against the urge to hold a grudge. Fight against the lie that you are better off on your own. Refuse to be cut off. Reach out to the soldier next to you. Reinforce the line. A soldier alone is a soldier defeated.

Stand **together**.

Chapter 17

Counterinsurgency
(Suppressing Our Will to Resist)

Military Warfare: The War for the Will

WHEN THE SHOCK and awe of conventional war subsides and a territory is officially "conquered," a new and far more complex type of conflict often emerges from the shadows. The occupying army may have defeated the enemy's uniformed divisions and captured its capital, but it now faces a persistent, asymmetrical, and often brutal resistance from within the population itself. This is the war *after* the war. This is insurgency: a rebellion that wages war not with tank columns and fighter jets, but with ambushes, assassinations, propaganda, and the unwavering conviction that they are fighting for their homeland. The sophisticated, multifaceted effort to defeat this type of rebellion is known as Counterinsurgency, or COIN.[1] It is not a battle for ground, which is already held; it is a grinding, soul-crushing war for the will of the people.

Counterinsurgency is one of the most difficult, costly, and psychologically demanding forms of warfare ever conceived. The traditional metrics of success – territory seized, enemy soldiers killed – become dangerously misleading.[2] The true objective of COIN is to make the insurgency irrelevant by systematically dismantling its support structures and, most critically, by crushing the insurgents' will to continue the fight. This is achieved through a two-pronged strategy, an "iron fist" of security wrapped in the "velvet glove" of political and social action.[3]

The first prong is establishing **dominating security and control.**[4] The occupying power must prove that it is in charge and that resistance is both dangerous and futile. This involves a heavy and pervasive intelligence operation to identify, track, and disrupt insurgent cells. It means constant patrols, random checkpoints, and swift, decisive raids on suspected safe houses. This physical suppression – the "iron fist" – is designed to break the logistical back of the insurgency and make every day a struggle for survival for its members.

The second, and ultimately more critical, prong is the battle for **legitimacy and popular support.**[5] This is the "velvet glove," the sophisticated psychological operations (PSYOP) campaign to win the "hearts and minds" of the population. The occupying force must convince the populace that its rule is more stable, just, and beneficial than the chaotic violence offered by the rebels. This involves a massive propaganda effort to discredit the insurgents, painting them as criminals or foreign-backed terrorists who only bring suffering. Simultaneously, the COIN force works to provide tangible benefits: rebuilding infrastructure, restoring electricity, distributing food and medical aid, and establishing a new, functioning local government that appears to serve the people's interests.

Case Study: The Roman Occupation of Judea

This brutal logic is not a modern invention; it was perfected by the ancient masters of occupation: the Romans. Following the First Jewish-Roman War (66-73 AD), which culminated in the destruction of the Temple in Jerusalem, Rome faced a simmering insurgency in the province of Judea.[6] Their COIN strategy was methodical and ruthless. They established a permanent and overwhelming military presence (the "iron fist"), garrisoning the Tenth Legion in the ruins of Jerusalem to project constant power. They imposed heavy taxes and appointed loyal local leaders to manage the population, creating a system of control. When a second, more widespread rebellion erupted under Simon bar Kokhba (132-136 AD), the Roman response was a campaign of annihilation designed to crush the will to resist for all time.[7] Roman legions systematically isolated rebel-held towns and slaughtered their inhabitants. Following their victory, Emperor Hadrian enacted the "velvet glove" of Romanization with extreme prejudice. He outlawed circumcision, renamed Jerusalem "Aelia Capitolina," built a temple to Jupiter on the Temple Mount, and renamed the province "Syria Palaestina." The goal was not just to defeat the rebels, but to erase the very cultural and religious identity that fueled their rebellion.

Case Study: The Malayan Emergency (1948-1960)

A more modern and strategically nuanced example of a successful COIN campaign is the Malayan Emergency.[8] The British faced a well-organized communist insurgency determined to establish a communist state. Realizing that purely military action was failing, the British under General Sir Harold Briggs implemented a comprehensive COIN plan, named "The Briggs Plan," designed to isolate the insurgents from their support base. The "iron fist" involved the forced resettlement of over 500,000 rural civilians – mostly ethnic Chinese who were the insurgency's primary source of food, intelligence, and recruits – into fortified "New Villages."[9] These villages were surrounded by barbed wire and heavily patrolled, physically cutting the insurgents off from the population. A strict food rationing system was imposed to starve the rebels in the jungle. Meanwhile, the "velvet glove" was applied. The British promised eventual independence for Malaya, offering the general population a more appealing future than the one offered by the communists. They invested in the infrastructure of the New Villages and granted the inhabitants land ownership, giving them a tangible stake in the government's success. This combined strategy methodically drained the "sea" of popular support, leaving the insurgent "fish" exposed and easy to hunt. The Malayan Emergency demonstrates the core principle of COIN: you defeat the insurgency by taking away its reason for being and its ability to survive, ultimately convincing its fighters that their cause is hopeless.

Spiritual Warfare: The War for the Will

When an enemy has successfully established a stronghold within a believer's life (Chapter 15) and has begun to isolate them from their allies (Chapter 16), the final stage of the siege begins. This is not a battle for new ground; it is a war for the will. This is a spiritual counterinsurgency campaign, a sophisticated psychological operation designed to crush the believer's innate, God-given spirit of resistance. The enemy, now an occupying force, seeks to convince the "insurgent" spirit that the fight for freedom is over, that resistance is futile, and that the new, compromised reality is permanent.

This is the most insidious phase of the siege. The enemy's goal is to move you from being a prisoner of war who still dreams of escape to a compliant subject who has accepted their chains. He aims to extinguish the very hope that fuels rebellion. He does this not with a single, overwhelming attack, but with a grinding, multi-pronged PSYOP campaign designed to demoralize you into total submission.

1. The Propaganda of Hopelessness (Crushing Morale): The primary weapon in any COIN campaign is the propagation of hopelessness. Satan bombards the mind with a relentless narrative of defeat. He whispers deceptive, weaponized lies designed to make you believe that change is impossible.

o *"You've tried to beat this before and failed every time. What makes you think this time will be different?"*
o *"This is just who you are now; you will never be free from this."*
o *"Look how far you've fallen. God is done with you; He's moved on to more faithful soldiers."*

This is a direct assault on the future, designed to convince you that your struggle is pointless and your destiny is defeat. The enemy wants to create a crushing sense of inevitability, where surrender feels not like a choice, but like a simple acceptance of the facts.

The Twin Weapons of Guilt and Shame (Attacking Past and Identity): The enemy masterfully exploits past failures to keep you imprisoned in the present. He uses two distinct but related weapons:

o **Guilt:** This is the weapon that keeps you focused on your *actions*. He will constantly replay the tape of your sin, forcing you to relive the failure, the pain you caused, and the commands you broke. The goal of guilt is to keep you staring at your sin so that you are too distracted to look at your Savior.

o **Shame:** This is a deeper, more venomous attack that targets your very *identity*. While guilt says, "You *did* something bad," shame says, "You *are* bad." It takes the sin and tries to make it the core of who you are. The enemy uses shame to make you feel fundamentally flawed, dirty, and unworthy of love or forgiveness. It is the lie that your failure has permanently disqualified you from your identity as a child of God. This is why you must cling to the truth of Romans 8:1: "Therefore, there is now no condemnation for those who are in Christ Jesus." This verse is a direct counterattack against the enemy's campaign of shame.

The Deployment of Self-Condemnation (Turning You Against Yourself): A successful COIN campaign convinces the occupied population to police itself. Spiritually, Satan achieves this through self-condemnation. He whispers his accusations for so long that you eventually adopt his voice as your own inner monologue. You become your own harshest critic, your own accuser. You stop needing him to condemn you because you are doing it for him. This is a tactic that erodes your self-worth, blinds you to the grace of God, and makes you an active participant in your own suppression. You begin to treat yourself with a harshness that God never would, rejecting the compassion and kindness that are central to His nature.

The Chemical Warfare of Spiritual Apathy (Inducing Numbness): If the propaganda of hopelessness and the attacks of shame fail to secure a full surrender, the enemy resorts to a form of spiritual chemical warfare: apathy. This is a slow-acting poison designed to numb your spirit. After repeated cycles of sin and failure, the pain of conviction can become exhausting. The enemy offers a satanic "peace" – a spiritual numbness where the sin no longer hurts as much. The conviction of the Holy Spirit, which once felt like a sharp, life-saving pain, now feels like a dull, distant ache that is easy to ignore. This is a profoundly dangerous state. A soldier who no longer feels the pain of their wounds is a soldier who is bleeding out and doesn't know it. This spiritual desensitization makes what was once unacceptable now tolerable, and what was once tolerable now comfortable. It is the final stage of a successful COIN campaign, where the will to resist has not only been broken, but forgotten.

This entire campaign is a direct assault on the promise of 1 John 1:9, which guarantees that "If we confess our sins, he is faithful and just and will forgive us our sins and purify us from all unrighteousness." The enemy's COIN strategy is designed to make you believe that this promise is no longer for you. It is a war for your will, and it can only be won by defiantly clinging to the hope and truth that your Commander has already provided for your total liberation.

Counterintelligence: Fueling the Insurgency

To counter a spiritual COIN campaign, you must think like an insurgent. You must become the leader of the rebellion in your own occupied soul. When an enemy's entire strategy is to crush your will to resist, your most defiant act, your primary counter-offensive, is to stubbornly, relentlessly, and strategically *refuse* to be pacified. You must reject their propaganda, disrupt their control, and wage a persistent, asymmetrical war against their occupation. This is not a battle you fight when you feel strong; this is the desperate, daily fight you wage from within the enemy's own territory to keep the flame of freedom from being extinguished.

Satan's COIN campaign is designed to make you accept your chains. Your counterintelligence, therefore, must be a precise rebellion, fueled by divine hope and executed with holy defiance. The following directives are your field manual for keeping the insurgency alive.

Directive 1: Launch a Counter-Propaganda Offensive (Renewing the Mind)

The enemy's primary COIN weapon is a PSYOP campaign of lies designed to create hopelessness. Your counterattack must be to seize control of the narrative within your own mind. You must become a ruthless counter propaganda officer, challenge every enemy broadcast, and replace it with the truth from High Command.

Identify the Enemy Frequency: Learn to recognize the enemy's broadcast. It is always characterized by fear, anxiety, shame, guilt, and condemnation. When a thought enters your mind that says, "You'll never change," or "You're a failure," you must immediately identify it as enemy propaganda. Do not treat it as your own inner voice.

Jam the Signal with Truth: Your primary weapon against this propaganda is Scripture. The lie of shame is jammed by the truth of Romans 8:1: "Therefore, there is now no condemnation for those who are in Christ Jesus." The lie of hopelessness is jammed by the truth of Jeremiah 29:11, which declares that God has plans to give you "hope and a future." You must take these truths and speak them out loud, declaring them over your situation and using them to drown out the enemy's broadcast.

Broadcast on God's Frequency: It is not enough to simply reject lies; you must actively fill your mind with truth. This is the command of Philippians 4:8: "whatever is true, whatever is noble, whatever is right, whatever is pure, whatever is lovely, whatever is admirable – if anything is excellent or praiseworthy – think about such things." This is an active choice to change the channel, to deliberately focus on the goodness of God, His past faithfulness, and the beauty of His creation. This act of "renewing your mind" is a direct assault on the enemy's atmosphere of despair.

Directive 2: Break the Enemy's Intelligence Cycle (Confession and Vulnerability)

An occupying force maintains control through fear and secrecy. They want you to hide your failures in shame, because a shamed and isolated insurgent is an ineffective one. Your counter is to destroy their intelligence cycle through radical confession and vulnerability.

File an After-Action Report (Confession): The moment you fail or fall into the sin associated with the stronghold, your standing order is to run – not away from God in shame, but *towards* Him in confession. The promise of 1 John 1:9, "If we confess our sins, he is faithful and just and will forgive us our sins and purify us from all unrighteousness," is your operational guarantee. Confession is not about groveling; it is a strategic act. It brings the enemy's victory into the light, where the blood of Christ nullifies it. It robs him of the power of blackmail.

Link Up with Friendly Forces (Vulnerability): An insurgent alone is a target. Insurgents in a cell are a threat. You must resist the urge to hide. Break the enemy's encirclement by sharing your struggle with trusted allies – a pastor, a mentor, a small group. This single act of courage invalidates the enemy's lie that you are alone and that you would be rejected if people knew the truth. It brings in reinforcements and turns your individual battle into a collective one.

Directive 3: Execute Small-Scale Resistance (Building Momentum)

An insurgency rarely wins by capturing the capital in a single day. It wins through a thousand small acts of defiance that erode the enemy's control. When you feel too weak for a major battle, you must engage in small-scale resistance.

Practice Minor Sabotage: When you feel overwhelmed by apathy, choose one small act of obedience. Read one verse. Say one prayer of thanks. Send one encouraging text to a fellow soldier. These small actions are acts of sabotage against the enemy's campaign of numbness. They declare that your will is not yet broken.

Cultivate Defiant Gratitude: Gratitude is a weapon. When the enemy is broadcasting hopelessness, deliberately listing three things you are thankful for is an act of defiance. It forces your mind to acknowledge a reality different from the one the enemy is presenting, breaking the cycle of negativity.

These directives are not a formula for an easy victory. They are the gritty, daily tasks of a resistance fighter. They are the means by which you keep the insurgency alive, disrupt the enemy's occupation, and fight for the freedom that is already your birthright in Christ Jesus.

Biblical Assault: After-Action Reports on the War of Wills

Scripture provides us with raw, unfiltered intelligence on how the enemy wages his counterinsurgency campaigns. These are not just stories of failure; they are detailed debriefings of God's most formidable warriors facing sustained psychological assaults designed to break their will and pacify their spirits. In these accounts, we find the divine blueprint for fueling our own resistance, even from a position of profound weakness and defeat.

Case Study: The Apostle Paul's Thorn
(2 Corinthians 12) – The Insurgency of Weakness

If ever there was a high-value target for a demonic COIN campaign, it was the Apostle Paul. His effectiveness as a soldier for Christ was unparalleled. The enemy, unable to defeat him through conventional means like stoning, shipwreck, or imprisonment, launched a personal, persistent, and demoralizing insurgency against him.

The COIN Campaign: Paul identifies the agent of this campaign with tactical precision: a "thorn in my flesh, a messenger of Satan, to torment me" (2 Corinthians 12:7). This was not a random trial; it was a targeted operation. The Greek word for "torment" implies being struck with a fist – a relentless, bruising assault designed to wear him down, to humiliate him, and to drain his resolve. The objective was classic counterinsurgency: to make the fight so miserable and his own weakness so apparent that he would lose the will to continue.

The Counter-Offensive: Paul's initial response was logical: he repeatedly requested extraction, asking the Lord three times to remove the enemy agent. The answer he received from High Command fundamentally rewrote the rules of engagement for spiritual warfare. The Lord said, "My grace is sufficient for you, for my power is made perfect in weakness" (2 Corinthians 12:9). This was a divine counter strategy of the highest order. Instead of removing the source of the pressure, God provided a way to turn the enemy's attack into a source of strength. Paul learned to lead the insurgency *from* his weakness, not in spite of it. He declares his new battle doctrine: "Therefore I will boast all the more gladly about my weaknesses, so that Christ's power may rest on me... For when I am weak, then I am strong" (2 Corinthians 12:9-10). He took the enemy's primary weapon – his own weakness – and turned it into the very thing that allowed God's overwhelming power to flow through him. He refused to be demoralized by his weakness and instead made it the foundation of his victory.

Case Study: David's Debriefing
(Psalm 51) – Reenlisting After Catastrophic Defeat

After his catastrophic moral failure with Bathsheba, King David was a defeated man. He was guilty of adultery and murder. The enemy had achieved a total victory. Satan's COIN campaign in the aftermath would have been ruthless, bombarding David with a torrent of shame, self-condemnation, and the crushing lie that he was permanently disqualified from God's service. The goal was to take this broken king and pacify him into a state of silent, shame-filled retirement.

The Counter-Offensive: Psalm 51 is the after-action report of a soldier fighting to get back into the war and resisting the enemy's attempt to make defeat permanent.

Unflinching Confession: David does not minimize his failure. He confesses it with brutal honesty: "I know my transgressions, and my sin is always before me" (Psalm 51:3). This act of bringing the sin into the light immediately breaks the enemy's power of secrecy and blackmail.

Appeal to God's Character: He doesn't appeal to his own merit, which is gone. He appeals to God's "unfailing love" and "great compassion" (Psalm 51:1), basing his hope for restoration on the nature of his Commander, not himself.

The Plea for a Renewed Will: This is the core of his insurgency. He does not just ask for forgiveness. He asks for the strength to fight again. "Create in me a pure heart, O God, and renew a steadfast spirit within me... Restore to me the joy of your salvation and grant me a willing spirit, to sustain me" (Psalm 51:10, 12). He is asking for his *will to resist* to be rebuilt from the ground up. He is refusing to accept his compromised state as his new identity.

These case studies reveal a profound truth: the will to resist is not born from our own strength or moral perfection. It is a defiant fire fueled by a stubborn hope in God's grace – His power to work through our weakness and His mercy to restore us after our failures. It is the refusal to surrender to the enemy's narrative of defeat.

Closing Charge: Keep the Insurgency Alive

You have been shown the enemy's final move in the siege: the war for your will. Having established a stronghold and cut you off from support, his objective now shifts from invasion to occupation. His goal is to pacify the territory of your soul, to crush your spirit of resistance with a relentless counterinsurgency campaign of shame, hopelessness, and self-condemnation.

He wants you to believe the fight is over. He wants you to accept your brokenness as your new identity. He wants you to lay down your arms not because you have been defeated by force, but because you have been defeated by despair. He wants you to mistake the pain of your struggle for the finality of your defeat.

But you have been given the counterintelligence. You know that a flicker of defiance is a fire he cannot extinguish. Your charge, therefore, is to become the leader of the insurgency in your own heart. When he broadcasts the lie of your permanent failure, you will wage a counter-propaganda war with the truth of God's restorative grace. When he attacks with the shame of your past, you will assault him with the reality of your forgiveness. When he tries to numb you with apathy, you will answer with a single, small act of defiant obedience.

Your resistance is not fueled by your own strength, but by the stubborn hope of the Gospel. It is the hope that declares, like Paul, that God's power is perfected in your weakness. It is the hope that cries out, like David, for a renewed spirit even after catastrophic failure. The enemy may have occupied territory, but he has not yet won the war for your will. Do not let him. Fight for every inch.

Keep the insurgency **alive.**

Phase IV

The Consolidation Phase
(Spiritual Slavery)

The long siege has **ended**.

THE SOUNDS OF battle have faded. The desperate cries of resistance have been silenced. They have not been replaced by the celebration of victory, but by the chilling, orderly silence of total occupation. The enemy is no longer at the gates; he is on the throne. The flag that flies over the fortress of your soul has been changed.

This is the Consolidation Phase. This is the war *after* the war, where the victor's primary objective is no longer to defeat an enemy, but to pacify a population. The enemy's goal now shifts from assault to administration. He is no longer fighting a war; he is building a new society in his own dark image, using the raw materials of a life that was once dedicated to God.

In this final, terrifying phase, you are no longer a soldier fighting a battle. You are a subject in an occupied territory. The will to resist has been systematically dismantled, replaced by a dull, resigned acceptance of the new regime. The enemy's campaign now focuses on solidifying his control, disarming any remaining pockets of spiritual life, and rebuilding your thoughts, habits, and desires until they perfectly reflect his own.

The chapters that follow are not a battle plan for you to execute. They are a damage report. They are a field commander's grim assessment of what happens when a fortress falls. They are a portrait of what it means to become a slave to the very enemy you were created to conquer.

At this critical juncture, the structure of this field manual will change. This is intentional. The sections that follow are no longer battle plans for an active soldier. A slave is not issued counterintelligence briefings. A disarmed man is not given assault objectives. Therefore, the **Counterintelligence** and **Biblical Assault** sections have been removed.

They are replaced by two new, stark assessments:

o **The Strategic Consequence:** A grim damage report detailing the outcome of living in this defeated state.

o **Case Study in Defeat:** A cautionary biography of a biblical figure who reached this point of defeat.

You will also notice the brevity of these chapters. The tone is colder. This is by design. The analysis is no longer a lengthy strategic lesson; it is a concise and urgent field dispatch from a lost battlefield. The purpose of this phase is not to equip you for a fight that, for the person in this state, has already been lost. It is to show you the full, terrifying reality of the enemy's victory condition, so that you will do everything in your power to never arrive here.

Pay attention.

This is a **warning**.

Chapter 18

Establishing Security (Controlling Our Thoughts and Actions)

Military Warfare:
The New Order of an Occupied Territory

WHEN AN OCCUPYING army has successfully crushed a rebellion, its mission undergoes a fundamental transformation. The objective is no longer to fight, but to govern. It must establish a comprehensive security kit designed to maintain absolute control and normalize its authority. This is not simply about preventing another uprising; it is about creating a new, predictable order where the occupier's power is the unquestioned law of the land. The sound of gunfire is replaced by the quiet, pervasive presence of the patrol.

The first priority is to establish physical and psychological control over the population. This involves implementing strict security measures that are visible and constant. Checkpoints are erected on major roads, not just to catch insurgents, but to remind every citizen that their movements are monitored. Curfews are imposed, restricting activity and creating a sense of limitation. A heavy intelligence presence is cultivated to ensure that dissent is identified before it can coalesce into a threat. The goal is to create an atmosphere of such pervasive surveillance that the very thought of resistance feels futile and dangerous.

Simultaneously, the occupying force works to legitimize its control. It makes public examples of any who dare to defy the new order, with punishments designed not only to remove the dissenter but to terrorize the population into compliance. The Roman practice of crucifixion along major roads was a brutal but effective form of this security doctrine. It was a constant, horrific reminder of the price of rebellion.[1] Over time, the occupier seeks to make its presence the new normal. The population becomes conditioned to the checkpoints, accustomed to the curfews, and learns to police its own conversations for fear of the informant.

Case Study: The Stasi and the Perfection of Psychological Control

Perhaps no regime in modern history perfected this apparatus of internal security more chillingly than the German Democratic Republic (East Germany) through its Ministry for State Security, universally known as the Stasi.[2] Following its establishment under Soviet occupation, the Stasi's official motto was "Shield and Sword of the Party." Its mission, however, went far beyond traditional state security. Its goal was not merely to react to dissent, but to create a society so thoroughly permeated by surveillance that dissent could never take root in the first place.

The Stasi's "iron fist" was its vast network of informants, the *Inoffizielle Mitarbeiter* (unofficial collaborators). At its peak, it is estimated that one in every 63 East Germans collaborated with the Stasi in some capacity.[3] This network was its true genius. The Stasi didn't need to place an agent in every apartment building because it successfully turned neighbors against neighbors, children against parents, and husbands against wives. The psychological effect was devastating. No one could be sure who was listening. A careless word at work, a joke told to a friend, or a complaint whispered to a spouse could lead to a visit from the authorities, loss of a job, or imprisonment.

This created a state of societal paranoia where the population began to police itself. The Stasi perfected a method called *Zersetzung* (literally "decomposition" or "corrosion"), a form of psychological warfare designed to destroy a person's life without leaving physical scars.[4] Agents would break into a target's apartment and subtly rearrange furniture, reset clocks, or replace one type of tea with another, all to make the person believe they were going insane. They would spread malicious rumors at the target's workplace, destroy their professional reputation, and sabotage their personal relationships. The goal was to isolate the individual and shatter their mental stability, neutralizing them as a threat without the need for a public trial. The Stasi mastered the art of making the occupied population an active participant in its own suppression, a grim testament to the power of establishing total security by first conquering the mind.

Spiritual Warfare:
The Prison of Predictable Sin

When the spiritual insurgency of a believer's conscience has been crushed, the enemy begins his consolidation phase. His first objective is to establish "security" – to take control of our thoughts and actions, transforming what was once a chaotic battlefield into a predictable, manageable prison of sin. He is no longer fighting our resistance; he is now administering our slavery.

This is the state where a believer is no longer actively *fighting* a particular sin but is now passively *managing* it. The war for freedom has been abandoned, replaced by an attempt to make the slavery more comfortable. This is a terrifyingly common state, observable in the lives of those around us. Psychologically, it mirrors the concept of **"learned helplessness,"** a condition first identified by psychologist Martin Seligman, where a subject, after enduring repeated adverse events beyond their control, stops trying to avoid them, even when opportunities for escape are presented.[5] Spiritually, after repeated failures, the believer simply stops trying to escape their sin. They have learned to be helpless.

The Bible provides a chilling case study of this process in the life of King Saul. After being confronted for his disobedience, Saul did not engage in a long, drawn-out war against his own pride and jealousy. Instead, he began to *manage* it. His sin became predictable. He would feel convicted, repent superficially, and then return to his plots against David. He learned to live with his own corruption. This is the new, quiet order of a defeated soul. You can see Saul's pattern in lives today:

It's the man who no longer battles his addiction to pornography but now subconsciously schedules his life to ensure he has the private time to indulge it. He's not at war with his lust; he has negotiated a ceasefire where he serves the addiction in exchange for the temporary peace of its satisfaction.

It's the woman consumed by bitterness who, like Saul nursing his jealousy, no longer seeks reconciliation but actively looks for new offenses to nurse, because the feeling of righteous indignation has become a familiar, almost comforting, part of her identity. She protects her anger like a treasure.

It's the person trapped in anxiety who stops praying for deliverance and instead builds their entire life around avoiding situations that might trigger their fear. Their fear is no longer an invader; it is their commanding officer, dictating their movements and shrinking their world day by day.

In this state, the enemy establishes spiritual "checkpoints" in the mind. There are thoughts the believer learns they are not allowed to think. They cannot dwell on God's holiness because it brings an unbearable wave of shame. They cannot contemplate their calling or purpose because it brings a crushing weight of guilt over their compromise. Like the citizens of an occupied nation learning to avoid the patrols, they actively steer their own minds away from these "dangerous" spiritual topics. There is a way that appears to be right, but in the end, it leads to death (Proverbs 14:12).

The enemy's voice – the voice of the "accuser of our brothers and sisters" (Revelation 12:10) – becomes their own inner monologue. The lie is no longer an external suggestion; it is an accepted identity.

"This is just who I am. I'm just an anxious person. I just have a temper. I'll always struggle with this."

They have accepted the enemy's psychological assessment of them as fact, becoming a willing informant against their own freedom and forgetting the truth that if anyone is in Christ, they are a new creation (2 Corinthians 5:17).

The core truth of this state is found in Romans 6:16: "Don't you know that when you offer yourselves to someone as obedient slaves, you are slaves of the one you obey...?" The person in this phase has ceased their rebellion and is now offering a quiet, consistent obedience to their new master – be it lust, fear, anger, or despair. Their life is no longer a struggle for freedom; it is the managed, predictable, and secure routine of a slave.

The Strategic Consequence: The Death of a Soldier

The consequence of living under the enemy's established security is not merely a life of repeated sin; it is a slow, methodical death. When you offer yourself as a slave to sin, the Bible is clear about the wages you will be paid: *death* (Romans 6:16). This is not just a future physical event; it is a present spiritual reality. The enemy's objective has always been to "steal, kill, and destroy" (John 10:10), and in this phase, you witness his mission accomplished in the quiet devastation of your own life.

Theft: The first casualty is the theft of your spiritual vitality. Your **joy** is stolen, replaced by the fleeting, hollow pleasure of your managed sin, which always leaves a residue of shame. Your **peace**

is stolen, replaced by a constant, low-grade anxiety – the unrest of a guilty conscience and the fear of being exposed. Your **purpose** is stolen. The grand, eternal mission of serving your King is abandoned for the small, exhausting, and self-centered mission of managing your addiction, nursing your bitterness, or placating your fear. You are no longer a soldier with a divine objective; you are a prisoner whose only purpose is to get through the day without disrupting the warden.

Murder: The next consequence is the murder of your most vital connections. The primary casualty is your **intimacy with God.** Prayer becomes a painful monologue of guilt, or it is abandoned entirely because you feel like a hypocrite. The Bible, once a source of life, now feels like a book of judgment, its pages highlighting your failure. The fellowship you once had with the Father is dead, replaced by a formal, distant acknowledgement of a God you no longer feel you can approach. Your **authentic relationships** with others are also murdered. A slave to a secret sin cannot be truly known. You are forced to live a life of deception, hiding the true state of your soul from your spouse, your friends, and your church. The vulnerability and honesty required for true Christian fellowship die, and you become an island of one.

Destruction: Finally, the enemy begins the work of total destruction. Your **testimony** is destroyed. Your life no longer points to the liberating power of Christ; it becomes a silent, cautionary tale of defeat. You cannot speak of freedom when you are living in a prison. Your credibility as a witness for the Gospel is neutralized. Even more terrifying is the destruction of your **spiritual senses.** Over time, the constant, low-grade sin desensitizes you. The conviction of the Holy Spirit, which once felt like a sharp, life-saving alarm, fades to a dull, ignorable hum. You are, as 2 Corinthians 4:4 warns, being spiritually blinded. You lose the ability to discern truth from the enemy's lies because you have made his primary lie – your own defeat – the operating system of your life. This is the ultimate strategic consequence: a soldier who cannot see, cannot hear, and cannot feel the battle raging around him because, for all intents and purposes, he is already dead.

Case Study in Defeat: The Predictable prison of King Saul (1 Samuel)

To understand the grim reality of a life lived under the enemy's security, we must turn to the tragic after-action report of Israel's first king. The life of Saul, after his rejection by God, becomes a harrowing case study in what happens when a man stops fighting his sin and begins to administer it. He is the blueprint for a soul under demonic occupation, a man whose thoughts and actions became so predictable that he was no longer a king, but a slave in a crown.

Saul's insurgency against God was definitively crushed in his disobedience regarding the Amalekites (1 Samuel 15). From that moment, the enemy did not need to launch new, creative assaults; he simply had to help Saul establish a "secure" and predictable state built around his primary sin: jealous pride. Saul's life devolved into a managed routine of rebellion.

His jealousy toward David was not a fleeting emotion; it became the new law of his land, the security protocol by which all other decisions were made. His rages were no longer surprising ambushes; they were the predictable patrols of an occupying force in his heart.

The Routine of Violence: When "an evil spirit from God rushed upon Saul," his reaction was not to fight it, but to act it out. He took his spear and hurled it at David, not once, but repeatedly throughout his reign (1 Samuel 18:10-11, 19:9-10). This was not a moment of lost control; it was the predictable action of a man whose operating system was now murder.

The Management of Deceit: When direct violence failed, Saul switched to strategic deception, a key tactic in managing his sin. He offered David his daughter Michal in marriage, not as a blessing, but as a "snare," hoping the Philistines would kill him and save Saul the trouble (1 Samuel 18:21). This was not a desperate gamble; it was a calculated move administered by a heart fully under the enemy's control.

The most telling evidence of Saul's new "secure" state was the complete pacification of his conscience. He could go through the motions of remorse, as seen when David spared his life in the cave, weeping and declaring, "You are more righteous than I" (1 Samuel 24:17). But this was a temporary ceasefire with his conscience, not true repentance. The security apparatus of his pride and jealousy was too deeply entrenched. Shortly after, he was back to hunting David again (1 Samuel 26).

The final, tragic proof of his subjugation came at Endor (1 Samuel 28). Having been completely cut off from his own Commander – "the LORD did not answer him" – Saul, in desperation, sought intelligence from the enemy. He consulted a medium, an act explicitly forbidden by the law he was supposed to uphold. This was the act of a defeated and fully occupied leader, so controlled by the enemy's security apparatus that he willingly turned to the demonic realm for guidance. He was a king who had become a slave to fear, a man whose thoughts and actions were the predictable, destructive patterns of a life lived under the full security of the enemy.

Closing Charge: The Quiet Prison

The damage report is filed. You have been shown what happens when the insurgency of the heart is crushed. The enemy no longer needs to fight you, because you have learned to police yourself. He does not need to shout his lies, because you now whisper them to yourself.

Look closely at this state. There is no struggle. There is no war. There is only the quiet, predictable, and soul-crushing routine of managed sin. The enemy has established his security by convincing you to accept your prison bars as the new shape of your world. He has won, not by a final, glorious battle, but by your slow, silent, and exhausted surrender.

This is not a battle. This is an **occupation**.

Chapter 19

Disarmament and Demobilization (Stripping Us of Our Spiritual Weapons)

Military Warfare:
The Neutralization of a Defeated Force

A CONQUERED POPULATION, even one whose rebellion has been crushed, still possesses the tools, the training, and the cultural memory of resistance. To secure a lasting and total occupation, the victor must engage in the systematic process of **disarmament and demobilization.**[1] This is not merely about collecting rifles; it is a profound psychological and structural campaign designed to dismantle the enemy's entire capacity to ever wage war again. It is the methodical erasure of their identity as a fighting force, ensuring a permanent and undisputed peace on the victor's terms.

The process is two-fold. Disarmament is the most visible component: the confiscation of personal weapons, the seizure of artillery, the scrapping of tanks and aircraft, and the shuttering of factories capable of producing arms.[2] Demobilization is a deeper, more structural act. It is the formal disbanding of the defeated army's command structure, the dissolving of its historic units, and the forced re-integration of its soldiers into civilian life, stripped of their rank and military purpose.[3] The ultimate goal is to leave the conquered people not only without weapons but without the organization, the leadership, or the cultural identity of a warrior class. A man with a rifle is a threat; a nation of disarmed civilians is a manageable population.

Case Study: The Post-War Demilitarization of Japan

Following Japan's unconditional surrender in World War II, the American-led occupation under General Douglas MacArthur executed one of the most comprehensive disarmament and demobilization campaigns in modern history.[4] The goal was not merely to prevent a future war but to fundamentally transform Japanese society from a militaristic empire into a peaceful nation. The "iron fist" of disarmament was swift and total. The massive Imperial Japanese Army and Navy were not just defeated; they were completely dissolved. Millions of soldiers were demobilized and returned to their homes.[5] All military factories, naval bases, and arsenals were systematically dismantled. In a move of immense symbolic power, traditional swords – potent icons of the samurai warrior spirit – were confiscated and destroyed, a direct assault on the nation's martial identity.

The masterstroke of this campaign, however, was psychological and constitutional. The new Japanese Constitution, drafted under the supervision of the occupying authority, included the now-famous Article 9, which states that "the Japanese people forever renounce war as a sovereign right of the nation" and that "land, sea, and air forces, as well as other war potential, will never be maintained."[6] This was the ultimate demobilization. It did not just take away their existing weapons; it took away their legal and moral right to ever possess them again. It was a strategic effort to disarm not just the hands of the soldiers but the mindset of the entire nation, creating a demilitarized state where the ability, the right, and even the desire to fight were, for a time, extinguished.

Spiritual Warfare:
The Laying Down of Arms

After an enemy has established security and controls the thoughts and actions of a defeated believer, his next strategic objective is **disarmament.** He must ensure that the subject of his occupation can never effectively fight back again. This is not a physical confiscation of weapons, but a sophisticated psychological campaign to convince the believer to lay down their own arms, rendering themselves completely defenseless. The enemy seeks to make us believe that our weapons are useless, that the fight is pointless, and that we are utterly powerless (2 Corinthians 10:3-4).

This spiritual demobilization is a slow and insidious process, a corrosion of the will to even reach for the weapons God has provided. You can see the tragic results in the lives of those who have been fully subjugated:

The Sword of the Spirit Becomes a Closed Book: The enemy's first target is the Word of God (Ephesians 6:17). He works to make the Bible a source of condemnation rather than life. For the believer trapped in managed sin, opening Scripture no longer brings comfort; it brings a painful awareness of their failure. The words of truth feel like accusations. Soon, they stop reading it altogether to avoid the guilt. The sword is not stolen; it is willingly left on the nightstand to gather dust, too heavy and too painful to lift.

The Shield of Faith Erodes into Opinion: Faith, the shield meant to "extinguish all the flaming arrows of the evil one" (Ephesians 6:16), is systematically undermined. The enemy bombards the believer with so many failures and unanswered prayers that their vibrant, trusting faith in a good and powerful God devolves into mere intellectual assent. They "believe" in God historically, but they no longer trust Him practically. The shield is laid aside, and the enemy's arrows of doubt, fear, and accusation now find their mark with devastating regularity.

Prayer Becomes a Formalized Surrender: The direct line to High Command is severed not by an enemy attack, but by the believer's own shame. Prayer, once a confident dialogue with the Commander, becomes a series of guilty, muttered apologies or is abandoned entirely. The enemy convinces the believer that they are too unworthy to approach God (Luke 8:14), turning the act of prayer from a weapon of warfare into a painful reminder of their defeated state.

The Belt of Truth and Breastplate of Righteousness are Removed: The truth of God's Word is replaced by the "truth" of the believer's feelings and failures. Their righteousness is no longer seen as a gift from Christ (2 Corinthians 5:21) but as something they have failed to earn. They live in a state of self-condemnation, and so they walk without the protection that comes from knowing who they truly are in Him.

This is the state of a disarmed soldier. They are a living fulfillment of the enemy's primary objective: to render a child of God utterly defenseless, not by force, but by convincing them to willingly participate in their own demobilization. They are left standing on the battlefield, facing a relentless enemy with no weapon in hand and no armor to protect them.

The Strategic Consequence: The Unarmed Man in the Kill Zone

The strategic consequence of spiritual disarmament is not a state of peaceful retirement, but one of perpetual, undefended terror. A soldier who has been convinced to lay down his arms has not found neutrality; he has become a walking target. He is an unarmed man standing in the middle of a kill zone, utterly exposed to an enemy who has no intention of holding his fire. This is the horrifying reality of a demobilized believer.

Total Vulnerability: Without the shield of faith, you are left defenseless against the "flaming arrows of the evil one" (Ephesians 6:16). Every accusation, every whisper of doubt, every wave of fear hits its mark with full, unmitigated force. There is nothing to deflect the blows. You are left to absorb the full psychological and spiritual trauma of the enemy's assault, leading to a state of constant anxiety and spiritual torment.

Complete Combat Ineffectiveness: A disarmed soldier cannot fight. You are not only unable to defend yourself, but you are also completely incapable of going on the offensive. You cannot reclaim territory from the enemy, you cannot fight for the freedom of others, and you cannot advance the cause of your King. You are rendered completely combat-ineffective, a non-participant in the great spiritual war, neutralized not by a mortal wound, but by your own compliance.

The Erasure of Identity: A soldier's identity is inextricably linked to his purpose and his weapons. When you lay down the sword of the Spirit and the shield of faith, you do not just lose equipment; you lose your identity as a soldier of Christ. You cease to see yourself as a warrior in a cosmic battle and begin to see yourself as a victim of your circumstances. Your identity shifts from "soldier" to "slave," from "conqueror" to "conquered."

The Silence of Command: The most devastating consequence is the severance of your connection to High Command. Prayer and the Word are not just weapons; they are your primary means of communication with your Commander. When you stop using them, the line goes quiet. The silence is not because God has stopped speaking, but because you have turned off your radio. You are left truly and terrifyingly alone, without orders, without comfort, and without the reassuring voice of the One who promised to lead you to victory. This is the ultimate end-state of disarmament: a terrified, ineffective, and isolated soul, waiting for the enemy's final blow.

Case Study in Defeat:
The Demobilization of Samson

To witness the catastrophic outcome of spiritual disarmament, we must examine the after-action report of Samson. No figure in

Scripture provides a more visceral or heartbreaking example of a divinely empowered warrior who was rendered utterly defenseless, not by a superior enemy force, but by a slow, willing demobilization of his own spirit. Samson was a one-man army, a soldier of immense power, who was ultimately defeated because he voluntarily handed the enemy the schematics to his own weapons system.

Samson's disarmament began long before Delilah ever entered the picture. It started with a consistent pattern of compromise, a willing fraternization with the enemy that eroded his status as a consecrated warrior. As a Nazirite, his entire life was to be a weapon dedicated to God, but he repeatedly chose to entangle himself with the very Philistine enemy he was commissioned to fight (Judges 14:1-3). He treated his divine power not as a sacred trust, but as a personal superpower to be used for his own gratification. This casualness with his commission was the first step in laying down his arms.

His relationship with Delilah was the final stage of his demobilization. The enemy, unable to defeat him in open combat, resorted to intelligence operations. Delilah's task was simple: discover the source of his strength so he could be disarmed. The back-and-forth between them is a harrowing depiction of a soldier being psychologically worn down until he surrenders his most vital secret. Three times, Samson lied to her, and three times, he allowed the enemy to test his false intelligence (Judges 16:6-14). Yet he never recognized the clear and present danger. He played games with the enemy, foolishly believing he could control the situation.

Finally, "pestered to death by her nagging," Samson "told her everything" (Judges 16:16-17). He revealed the secret of his Nazirite vow – the source of his strength. This was the ultimate act of demobilization. He did not just give the enemy intelligence; he gave them the disarmament manual.

The result was immediate and catastrophic. After his hair was cut, Delilah cried out, "Samson, the Philistines are upon you!" The Bible then records one of the most terrifying verses in all of Scripture:

"He awoke from his sleep and thought, 'I'll go out as before and shake myself free.' **But he did not know that the LORD had left him.** " (Judges 16:20)

This is the state of the disarmed soldier. He still believes he has his power. He still goes through the familiar motions of warfare, but the divine connection is gone. He is a warrior in his own mind but a helpless man in reality. The Philistines seized him, gouged out his eyes, and put him to work grinding grain in a prison. The man who was once the judge of Israel, God's chosen weapon, was now a blind, powerless slave, a source of amusement for his captors. He was a fully demobilized soldier, a living testament to the fact that divine power, when treated with contempt, can be removed, leaving the once-mighty warrior utterly and shamefully defeated.

Closing Charge: The Unarmed Man

The Sword of the Spirit lies rusting on the ground. The Shield of Faith has been abandoned. The line to High Command is silent, not because the Commander isn't speaking, but because the radio has been willingly smashed.

He is not a veteran enjoying a peaceful retirement. He is a man standing bare in a kill zone, with nothing to offer but his own unprotected flesh.

You are no longer a soldier.

You are a **casualty**.

Chapter 20

Establishing a New Government (Replacing God's Authority with His Own)

Military Warfare:
The Installation of a Puppet Regime

ONCE AN OCCUPYING army has secured a territory and disarmed the populace, its most sophisticated act of control is not to rule directly, but to establish a new, seemingly legitimate government. This is the installation of a **puppet regime** – a government that maintains the outward appearance of sovereignty but is, in reality, completely controlled by the foreign occupier.[1] This strategy is a masterstroke of psychological and political warfare. It allows the victor to rule from the shadows, outsourcing the daily, often brutal, work of administration and oppression to local figures, thereby creating a façade of normalcy that pacifies the population and makes resistance seem like an act of treason against one's own people.

The puppet government serves several critical functions for the occupying power. It provides a veneer of legitimacy, making the occupation seem more like a political transition than a hostile takeover.[2] It leverages familiar faces and national symbols to encourage compliance. Most importantly, it co-opts local leaders, forcing them to become the public face of the occupier's policies. This deflects popular resentment away from the foreign power and onto the collaborating local officials, fracturing internal unity and making it difficult for a cohesive resistance movement to form. The ultimate goal is to have the conquered people governed, taxed, and policed by figures who look like them but serve the interests of an unseen master.

Case Study: The Vichy Regime in France

Following the swift defeat of France by Nazi Germany in 1940, the Germans did not immediately occupy the entire country.[3] Instead, they endorsed the creation of a new French government based in the spa town of Vichy, led by the aging World War I hero, Marshal Philippe Pétain. This was the quintessential puppet regime. The Vichy government, officially known as the French State, had its own leader, its own ministers, and flew the French flag. To many, it appeared to be a legitimate, albeit weak, French government navigating a terrible crisis.[4]

In reality, Vichy was a collaborationist state, existing at the pleasure of and in service to Nazi Germany. While maintaining a façade of autonomy, the Pétain government actively participated in the occupiers' agenda. It enacted antisemitic laws without direct German orders, rounded up foreign and French Jews for deportation to concentration camps, and created a paramilitary force, the *Milice,* to help the Gestapo hunt down members of the French Resistance.[5] By installing a government led by a revered French marshal, the Germans created a brilliant and cynical trap. It forced the French people into a crisis of loyalty, where fighting against the German occupation also meant fighting against the "official" government of France. Vichy stands as a grim testament to the strategic power of replacing a legitimate government with a puppet regime that forces a nation to participate in its own subjugation.

Spiritual Warfare:
The Puppet Government of Self

After the believer has been pacified and disarmed, the enemy executes his most brilliant and insidious political maneuver: he installs a new government on the throne of the heart. He knows that a direct, overt rule would be easily identified as tyranny. Instead, he establishes a **puppet regime.** And the leader he installs, the one who appears to be in charge and making all the decisions, is **you.**

He replaces God's authority not with his own demonic dictates, but with the supremacy of your own desires, passions, and intellect. The central question of your life is no longer, "What does God want?" but is now, "What do *I* want?" or "What feels right to me?" This is the essence of his new government. By convincing you to serve yourself, he ensures that you will never again serve God, and he rules from the shadows, content to let his puppet believe it is free.

This is the state where a person's life is no longer governed by the clear commands of Scripture but by the shifting whims of a new, internal authority.

The Government of Emotion: This regime makes feelings the ultimate arbiter of truth. Decisions are based not on God's will, but on what "feels good" or what will bring the most immediate emotional comfort. This government will sacrifice long-term holiness for short-term happiness, abandon difficult commitments when they become emotionally taxing, and justify any action as long as the "heart was in the right place."

The Government of Ambition: Here, the "pride of life" (1 John 2:16) is enthroned. Career goals, financial accumulation, social status, and personal reputation become the guiding principles of life. The believer may still use "Christian" language, but their major life decisions are made in the service of their own personal kingdom, not God's.

The Government of Intellect: This regime elevates human reason and cultural consensus above divine revelation. The clear teachings of the Bible on morality, sacrifice, or controversial social issues are dismissed as "outdated" or "problematic." The believer becomes their own theologian, creating a custom-made god who endorses their preferred lifestyle and political views.

In each case, the believer has become a living embodiment of Romans 1:25: "They exchanged the truth about God for a lie and **worshiped and served created things rather than the Creator.**" The "created thing" they now serve is their own self. By placing their own will on the throne, they have committed the ultimate act of treason, turning the province of their soul into a breakaway kingdom that, while flying the flag of personal freedom, ultimately serves the strategic interests of the great enemy of God.

The Strategic Consequence:
The Kingdom of Self

The strategic consequence of installing a puppet government of Self is a takeover in the soul. It is a fundamental act of treason where loyalty to the true King is replaced by service to a worthless usurper. The outcome is not freedom, but the establishment of a hollow, chaotic kingdom destined for ruin.

The Death of Sovereignty: The first casualty is the sovereignty of God in your life. He is demoted from absolute Monarch to a constitutional advisor, whose counsel can be accepted or ignored based on the whims of the new ruling power: your feelings, your ambitions, or your intellect. True, unconditional obedience dies. You no longer live to serve His will; you now expect Him to serve yours, to bless the plans you have made in service to yourself. Your life ceases to be a mission from a King and becomes a business plan for a president of one.

The Anarchy of a Failed State: The second casualty is peace. A government of Self is a government of chaos. Your emotions, a fickle and tyrannical ruler, plunge your inner world into civil war. Your ambitions, a paranoid and insatiable dictator, demand constant appeasement. Your intellect, an arrogant and elitist legislator, constantly rewrites moral law to suit its own comfort. The result is a life marked by internal conflict, anxiety, and a desperate, unending struggle to hold your fragile kingdom together. This isn't freedom; it's the anarchy of a failed state.

The Perversion of Worship: The final and most devastating consequence is the perversion of worship. Life becomes a constant, exhausting effort to appease the new deity on the throne: "Me." Every decision is filtered through the question of how it will benefit, gratify, or protect the self. You are now living out the grim reality of Romans 1:25, having "worshiped and served created things rather than the Creator." The created thing you now serve is the idol of your own reflection. The kingdom of your soul, once meant to be a glorious embassy of Heaven, has become a small, self-obsessed, and ultimately bankrupt dictatorship.

Case Study in Defeat:
The Divided Kingdom of Solomon

To witness the catastrophic end of a life governed by a puppet regime, we must examine the reign of King Solomon. No one in history began with a more legitimate, God-ordained government in his heart. When God offered him anything, Solomon asked only for "a discerning heart to govern your people and to distinguish between right and wrong" (1 Kings 3:9). God was his undisputed King, and Solomon ruled as His faithful viceroy. His early reign was the golden age of Israel, a kingdom of peace and wisdom under the direct authority of God.

But the consolidation of his own power led him to make treaties sealed by political marriages. He "made an alliance with Pharaoh king of Egypt and married his daughter" (1 Kings 3:1) and went on to amass hundreds of wives and concubines from foreign nations. These were not just personal relationships; they were the installation of foreign embassies in the capital of his heart. And with these foreign wives came their foreign gods.

The Scripture is explicit about the coup that followed: "As Solomon grew old, his wives turned his heart after other gods, and his heart was not fully devoted to the LORD his God, as the heart of David his father had been" (1 Kings 11:4). The legitimate King, Yahweh, was deposed. A new coalition government of foreign deities, championed by his wives, now held the real power. Solomon became a puppet king in his own soul, serving the interests of these new masters.

His actions became those of a collaborationist ruler. On a hill east of Jerusalem, the man who built the glorious Temple to the one true God now built a "high place for Chemosh, the detestable god of Moab, and for Molek, the detestable god of the Ammonites" (1 Kings 11:7). He was no longer acting as God's servant; he was appeasing the foreign powers he had allowed to set up shop in his kingdom and in his heart.

The verdict from the true King was swift and devastating. The LORD became angry with Solomon and declared: "Since this is your attitude and you have not kept my covenant and my decrees... I will most certainly tear the kingdom away from you" (1 Kings 11:11). Solomon began his reign with a heart so submitted to God that it brought him unparalleled wisdom and peace. He ended it as the ruler of a divided kingdom, a man whose great intellect was enslaved to the foolishness of idolatry, a king who had willingly handed the throne of his heart over to a puppet regime of foreign gods.

Closing Charge: The Puppet King

You have witnessed the overthrow of the soul. The throne is not empty. It has been claimed by a usurper who looks just like you. This new government promises freedom but delivers only the anarchy of a failed state, ruled by the chaos of your own desires. You are no longer a servant of the true King. You are the loyal subject of a puppet regime, and the enemy you cannot see delights in your treason.

This is not leadership. This is **idolatry**.

Chapter 21

Reconstruction and Development (Rebuilding Our Lives in Satan's Image)

Military Warfare:
The Architecture of a New Order

ONCE AN OCCUPYING power has established security and installed a compliant government, its final and most ambitious act of consolidation begins: **reconstruction and development.**[1] This is far more than simply rebuilding bombed-out cities; it is the strategic act of reshaping a defeated nation in the victor's own image. It involves rebuilding the country's political, economic, and social infrastructure according to the occupier's values and systems. The ultimate goal is to create a new reality, a self-sustaining state that is no longer an enemy, but a stable, predictable partner aligned with the victor's worldview.[2] This is the architecture of a lasting "peace", secured on the victor's terms.

This reconstruction is a comprehensive project. It often involves massive infusions of economic aid, but this aid always comes with strings attached.[3] The defeated nation must adopt new laws, new economic models, and new forms of government that mirror those of the occupier. Old industries may be broken up, new educational curricula are introduced, and cultural exchanges are promoted, all with the aim of slowly and methodically replacing the old national identity with a new one that is compatible with the victor's global interests.

Case Study: The Post-WWII Reconstruction of Germany and Japan

Following the total defeat of the Axis powers in 1945, the Allies, led by the United States, embarked on the most extensive reconstruction project in history. In Europe, the **Marshall Plan** poured billions of dollars into rebuilding Western Germany, but it did so to create a fortification against communism. The aid was designed to build a prosperous, democratic, and capitalist Germany that would be ideologically aligned with the West.[4]

In Japan, the American occupation under General Douglas MacArthur was even more direct. MacArthur oversaw a revolutionary "reconstruction" of Japanese society.[5] He forced the adoption of a new constitution (often called the "MacArthur Constitution"), which renounced war and established a parliamentary democracy. He implemented sweeping land reforms that broke the power of the old feudal landlords and created a new class of independent farmers. He dismantled the *zaibatsu,* the massive industrial conglomerates that had fueled the Japanese war machine.[6] The explicit goal of these reforms was to fundamentally change the nation's character, surgically removing the pillars of militarism and imperialism and replacing them with the democratic and capitalist values of the United States. This was not just rebuilding; it was a complete societal re-engineering, designed to turn a bitter enemy into a permanent and stable ally.

Spiritual Warfare:
The Life Built on the Wrong Foundation

Once the puppet government of Self is firmly in power, the enemy begins his grand "Reconstruction and Development" project. He is no longer content to simply manage your sin; his new objective is to rebuild your entire life into a functional, respectable, and even impressive monument to *his* values. This is not a project of demolition, but of perverse creation. He takes the raw materials of your life – your talents, your relationships, your career, your passions – and uses them to construct a life that is perfectly conformed to the pattern of this world.

The genius of this phase is its subtlety. The enemy does not rebuild your life around obvious, grotesque evil. Instead, he takes good, God-given things and expertly twists them into idols. He encourages you to build your identity, your security, and your purpose on these things rather than on Christ. This is the state where a person's life, from the outside, might look successful, moral, and even "blessed," but its entire foundation has been replaced.

Reconstruction around Career: The desire to work hard and provide for one's family is rebuilt into a relentless ambition where one's job title, income, and professional reputation become the source of their identity and worth. God is relegated to a silent partner in a business plan for personal glory.

Reconstruction around Family: The love for one's children is twisted into an idolatrous obsession where the children's success, happiness, and achievements become the parent's primary reason for living. The parent sacrifices their own spiritual life on the altar of their children's activities and future, finding their salvation in their children's acceptance to a good college rather than in Christ.

Reconstruction around Knowledge or Cause: The pursuit of knowledge or social justice is rebuilt into a fortress of intellectual pride. The person becomes so dedicated to their cause or their theological system that they lose all love and compassion for those who disagree. Their "rightness" becomes their righteousness, and they worship their own intellect rather than the God of all wisdom.

In each case, the enemy has successfully overseen the construction of a life that is a perfect reflection of "the lust of the flesh, the lust of the eyes, and the pride of life" (1 John 2:16). The person in this state is living in direct opposition to the command of Romans 12:2: "Do not conform to the pattern of this world, but be transformed by the renewing of your mind." The enemy has achieved the ultimate conformity. He has rebuilt their life into a beautiful, functional, and self-worshiping machine – a temple designed and built in the very image of Satan.

The Strategic Consequence: The Polished Tomb

The strategic consequence of the enemy's reconstruction project is the creation of his masterpiece: a beautifully functional, respectable, and utterly godless life. This is not a life of ruin and squalor, but a polished, impressive monument to the values of this world. The defeated believer becomes, in the chilling words of Jesus, a "whitewashed tomb, which looks beautiful on the outside but on the inside is full of the bones of the dead" (Matthew 23:27).

The Death of Transformation: The first casualty is the divine process of sanctification itself. The command to "be transformed by the renewing of your mind" (Romans 12:2) is halted and reversed. Instead of being made more like Christ, the person is expertly conformed to the pattern of the world. They become a perfect reflection of their culture's values: successful, driven, self-sufficient, and completely devoid of any true, desperate reliance on God. Their spiritual growth plateaus and then begins to erode as their energy is poured into building a life that *looks* good rather than a life that *is* good.

The Loss of Identity: The second casualty is your identity as a "new creation" (2 Corinthians 5:17). You are no longer a citizen of Heaven, a soldier of Christ, or a child of the King. You become defined by the very idols the enemy has used to rebuild your life. You replace those titles with a "successful lawyer," a "devoted mother," or a "brilliant academic." These titles, once aspects of your life, have now become the sum of your identity. Your worth is no longer found in the finished work of Christ, but in the unfinished work of your own résumé and reputation.

The hollowness of a counterfeit life: The final and most tragic consequence is the pervasive hollowness of it all. The life, for all its external success, is built on a foundation of sand. It offers no real peace, no lasting joy, and no eternal security. It is a beautiful house that cannot withstand the storms of tragedy, loss, or death. The person becomes an expert at maintaining the façade, but in the quiet moments, they are haunted by a profound sense of emptiness. They have gained a life that the world applauds but have lost the abundant life that only Christ can give (John 10:10). They are the living embodiment of a successful counterfeit, a masterpiece of the enemy's design.

Case Study in Defeat:
The Monument of King Ahab (1 Kings)

To witness the chilling reality of a life that has been entirely rebuilt in the enemy's image, we must examine the reign of King Ahab. While his father, Omri, built the kingdom of Israel physically, Ahab, through his monumental spiritual compromise, allowed the enemy to systematically reconstruct the nation's spiritual and cultural landscape, transforming it into a monument to foreign idolatry. His story is a devastating illustration of what happens when a leader, and by extension a people, undergoes "reconstruction and development" under the enemy's blueprint.

Ahab's downfall began not with a direct attack, but with a strategic alliance: his marriage to **Jezebel, the daughter of Ethbaal, King of the Sidonians** (1 Kings 16:31). This was far more than a political union; it was the ultimate act of cultural and spiritual reconstruction. Jezebel was not content to merely worship her own gods; she was an aggressive and zealous evangelist for Baal and Asherah. She became Satan's chief architect, overseeing the systematic dismantling of the worship of God, and the building up of Baal worship in Israel.

Rebuilding the Religious Infrastructure: Jezebel immediately set to work. "She built an altar for Baal in the temple of Baal that he had built in Samaria" (1 Kings 16:32). This wasn't just a personal shrine; it was a state-sponsored, publicly visible act of reconstruction. Prophets of Baal and Asherah, 850 of them, ate at her table (1 Kings 18:19), becoming the new official clergy, replacing the prophets of the Lord whom Jezebel systematically persecuted and murdered (1 Kings 18:4). This was a full-scale rebuilding of the religious system.

Reconstructing Cultural Values: Under Jezebel's influence, the nation's moral compass was dismantled. Justice was perverted, as seen in the story of Naboth's vineyard (1 Kings 21), where a man was framed and murdered so Ahab could seize his property. The reverence for God's law was replaced by lawlessness, and the pursuit of righteousness was eclipsed by the pursuit of land, power, and pagan debauchery.

Ahab was not merely passive; he became a willing participant in this grand reconstruction project. "He did more to arouse the anger of the LORD, the God of Israel, than all the kings of Israel before him" (1 Kings 16:33). He allowed his life, his reign, and his nation to be meticulously redesigned in the image of Baal worship—a religion characterized by fertility cults, self-mutilation, and child sacrifice. It was a life outwardly powerful, economically prosperous (built upon unrighteousness), and religiously vibrant (with Baal worship in full swing), but utterly corrupt at its core.

The consequences were devastating. God sent Elijah to confront this reconstruction, bringing a severe drought as a direct judgment against the false god of fertility (1 Kings 17). The showdown on Mount Carmel (1 Kings 18) was a direct confrontation between the true God and the false god that Ahab had allowed to reconstruct his nation. Though God powerfully demonstrated His sovereignty, Ahab's heart remained largely unmoved. He had become too deeply intertwined with the enemy's reconstruction.

Ahab's life ultimately became a monument to the enemy's values. He died in battle (1 Kings 22), not a glorious death, but one orchestrated by God's judgment. His legacy was one of profound spiritual compromise, a king who allowed his personal life and the nation he led to be expertly rebuilt according to the enemy's blueprints. He represents the terrifying success of the Satan's "reconstruction and development" plan: a life that, from the outside, appears to function and is beautiful, but on the inside is a hollow and godless tomb.

Closing Charge: The Monument

The reconstruction is **complete**. The enemy has taken the raw materials of a life once dedicated to God and has built his own temple. It is a polished and respectable structure, admired by the world. It is also a tomb.

You are no longer a soldier in the fight.

You are the monument to **his** victory.

Chapter 22

Winning Hearts and Minds (Blinding Us to the Truth)

Military Warfare:
The Campaign for Allegiance

THE FINAL, MOST sophisticated stage of consolidating power over an occupied territory is the campaign to **win hearts and minds.**[1] This is the strategic pivot from forced compliance to willing allegiance. An army can maintain control through the barrel of a gun for a time, but this is costly, inefficient, and breeds resentment that can fuel future rebellions. A truly successful occupation ends when the conquered population no longer sees itself as occupied, but as a willing partner or beneficiary of the new order.[2] Winning hearts and minds is the art of making the occupier's worldview seem not only normal but desirable.

This is a war fought not with bullets, but with propaganda, economic aid, and cultural influence. It is a deliberate campaign to change how a population thinks and feels. The occupying power may launch massive infrastructure projects, distribute food and medical supplies, and invest in the local economy, all to demonstrate that life is better under their rule. They will control the media and education systems to promote their own narrative, casting themselves as benevolent modernizers and any remaining resistance as backward-thinking terrorists. The goal is to create a generation that grows up knowing only the occupier's version of history and reality, making the new order feel like the only natural order.

Case Study: The Cold War's Cultural Battlefield

The global struggle between the United States and the Soviet Union was a textbook war for hearts and minds. Both superpowers understood that victory would not be won solely through military might but through ideological allegiance. They engaged in a massive, worldwide campaign to convince unaligned nations and even each other's populations that their system – capitalist democracy or communist collectivism – was the superior and more righteous path for humanity.

The United States Information Agency operated radio stations like Voice of America and Radio Free Europe, beaming news, rock-and-roll music, and cultural programming behind the Iron Curtain to showcase the freedom and prosperity of the West.[3] They sponsored international tours of jazz musicians and abstract expressionist art exhibits to highlight American cultural dynamism.[4] The Soviet Union countered with its own propaganda, funding communist parties worldwide, promoting its achievements in the space race as proof of its technological superiority, and holding up the West's history of colonialism and racial inequality as evidence of its moral bankruptcy.[5] This was not a war for territory in the traditional sense; it was a global competition to win the hearts and minds of the world, to convince billions of people that one way of life was inherently better than the other. It was a war for the allegiance of the soul.

Spiritual Warfare: The Contented Prisoner

After the enemy has pacified the soul, disarmed it, installed a puppet government of Self, and rebuilt its life around worldly values, he undertakes his final and most diabolical act of consolidation: the campaign to **win the heart and mind.** His objective is no longer to make you a compliant slave, but a *contented* one. He seeks to blind you to the truth of your own chains by making your prison cell so comfortable and well-lit with worldly pleasures that you no longer have any desire to leave.

This is a campaign of seduction, not terror. The enemy knows that a person can endure misery for a long time, but they will rarely abandon a life of comfort and camaraderie. He therefore offers counterfeit comforts, community, and purpose that feel more real and satisfying than the distant memory of God's love.

He offers counterfeit community. He knows we were created for fellowship. So, for the person who has left the church out of bitterness, he offers the passionate camaraderie of a political movement or a social cause. For the intellectual who has abandoned faith, he provides a smug community of skeptics who affirm their superiority. For the person enslaved to a hobby or lifestyle, he offers a vibrant subculture that celebrates their choices. In each case, he provides a sense of belonging and acceptance that satisfies the innate human need for connection, making the fellowship of the saints seem judgmental and dull by comparison.

He offers counterfeit purpose. He replaces the grand, eternal purpose of serving God's kingdom with smaller, more tangible, and self-aggrandizing missions. Your purpose becomes achieving the next career milestone, perfecting your body, raising "successful" children, or becoming an expert in your chosen field. These pursuits provide a sense of accomplishment and identity that masks the deeper worthlessness of a life disconnected from its Creator.

He offers counterfeit light. He blinds us by surrounding us with other, more appealing lights. The glow of the smartphone, the neon of entertainment, the flash of material wealth, and the warmth of public praise become so bright and constant that the pure, simple light of the Gospel seems faint and irrelevant.

This is the state described in 2 Corinthians 4:4: "The god of this age has blinded the minds of unbelievers, so that they cannot see the light of the gospel that displays the glory of Christ." The blinding is not an act of putting out their eyes; it is an act of distraction. The glory of Christ is eclipsed by the more immediate and tangible glories of the world. The person in this state is not miserable in their darkness; they believe they are living in the light. They have been won over. They are no longer a prisoner of war dreaming of home; they are a naturalized citizen of Satan's kingdom, content with their new life and blind to the truth that they have traded eternal glory for a comfortable cage.

The Strategic Consequence: The Kingdom of the Blind

The strategic consequence of the enemy's successful "hearts and minds" campaign is the final, most terrifying victory: he makes you an active, willing citizen of his kingdom of darkness. This is not just slavery; it is a state of such profound spiritual blindness that you no longer see the prison bars and have come to love the comfort of your cell.

The Death of Discernment: The first casualty is your ability to distinguish light from darkness. When you have been blinded by the enemy's counterfeit comforts and purpose, the true Light of the Gospel becomes painful. Truth feels like an attack. The conviction of the Holy Spirit feels like judgment. The loving correction of a fellow believer feels like a personal assault. Your moral and spiritual compass is not just damaged; it is inverted. You begin to call evil good and good evil, defending your bondage and becoming hostile to the very people who would seek to set you free.

The Final Atrophy: The spiritual "muscles" of faith, hope, and love for God, having been unused for so long, completely atrophy. There is no desire to pray, no hunger for the Word, no longing for true fellowship. These things are not just absent; they are alien. The soul, once created for intimate communion with God, is now fully satisfied by worldly pleasures and affirmations.

The believer is spiritually paralyzed but feels no distress, because the enemy has filled their life with so much noise, activity, and distraction that they never notice the silence of God.

The Great Exchange: The ultimate consequence is the willing exchange of eternal glory for temporary comfort. You have traded the challenging, life-giving truth of Christ for a comfortable, manageable lie. You have become a living embodiment of 2 Corinthians 4:4, a mind so thoroughly blinded by the god of this age that the light of the Gospel is not only unseen, but unwanted. This is the enemy's endgame: a soul that is not just lost, but content in its lostness, a prisoner who has been convinced that he is king of his own small, dark, and utterly hopeless world.

Case Study in Defeat:
The Blindness of the Pharisees

To understand the terrifying end-state of a successful "hearts and minds" campaign, we must look at the Pharisees during the ministry of Jesus. They are the ultimate case study in a group so completely won over by their own system that they became blind to the Truth Incarnate standing in their midst. They are a warning that the most profound spiritual blindness often wears the mask of religious devotion.

The Pharisees had, in effect, run their own successful hearts and minds campaign in Israel for generations. They had built an impressive, comprehensive system of righteousness based on the meticulous observance of the Law and their own traditions. They controlled the religious education, held the respect of the common people, and offered a clear (though burdensome) path to piety. Their lives were outwardly moral, their theology was rigorously debated, and their devotion was publicly displayed. They had successfully won the hearts and minds of the nation – and, most tragically, their own – for a kingdom of their own making.

Then, Jesus arrived. He did not come as a competing religious leader trying to build a better system. He came as Truth itself, and

His very presence was a threat to their entire world. His miracles were a direct challenge to their authority. His teachings of grace and mercy undermined their system of works-based righteousness. He offered freedom, which was terrifying to men who had built their identities on the careful administration of a prison.

Their blindness was not from a lack of evidence, but from a calculated choice. To accept Jesus as the Messiah would have required them to dismantle their entire kingdom. They would have had to give up their social standing, their political influence, and their spiritual pride. Their hearts and minds had been so thoroughly "won" by their own system that they were incapable of seeing the one true God. Jesus diagnosed their condition with surgical precision: "How can you believe since you accept glory from one another but do not seek the glory that comes from the only God?" (John 5:44).

This culminated in the tragic scene of John 9, where Jesus heals a man born blind. The man can now physically see, but the Pharisees, when confronted with this undeniable miracle, are spiritually blind. They interrogate the man, insult him, and throw him out of the synagogue for bearing witness to the truth. They loved the comfort of their system more than the disruptive power of a miracle.

In the end, the Pharisees became the ultimate cautionary tale. In their effort to protect the comfortable kingdom of darkness they had built, they conspired to kill the very Source of Light. They were prisoners who loved their cell so much that they murdered the one who held the key. They are the final, grim proof that the most effective way the enemy blinds a soul is by convincing it that it already sees perfectly.

Closing Charge: The Loyal Subject

The campaign is over. Satan no longer needs to force your compliance; he has won your allegiance. He has blinded you with a counterfeit light so brilliant that the true Light now seems like a shadow. You are not a prisoner crying out for freedom. You are a loyal subject, content in a kingdom of lies, and you call your blindness "sight."

This is not defeat.

This is **conversion**.

Epilogue

The Ultimate Rally Point

YOU HAVE JOURNEYED through the battlefield. You have meticulously studied the enemy's playbook – his subtle reconnaissance and strategic assessment in the pre-conflict phase, his rapid, decisive assaults and information warfare in the initial offensive. You endured the grinding attrition, the strategic isolation, and the relentless counterinsurgency of his long siege. And, most chillingly, you walked through the valley of the shadow of spiritual defeat in the consolidation phase – a landscape of predictable prisons, disarmed spirits, puppet governments, lives meticulously rebuilt in his image, culminating in a profound, self-imposed blindness.

This was not a pleasant journey. It was a stark, unvarnished look at the enemy's ultimate desired end-state for every believer: to steal, kill, and destroy. It was a necessary warning, designed to show you the terrifying consequence of failing to engage in spiritual warfare, of mistaking complacency for peace, or compromise for coexistence. The enemy's strategies are patient, cunning, and designed to lead to your utter subjugation. The grim reality of a life surrendered to spiritual slavery, where the chains are self-imposed and the prison comfortable, is a fate you must now understand at the deepest level.

But let us be unmistakably clear: **This war is already won**. The outcome is not in doubt.

The enemy you have so diligently studied, the one whose tactics have been laid bare across these pages, is a defeated foe. Two millennia ago, on a brutal hill outside Jerusalem, our Commander-in-Chief, Jesus Christ, launched the ultimate invasion. He did not merely win a battle; He executed the decisive, blow that broke the enemy's back and stripped him of his authority. The Cross was not a defeat for our King; it was the ultimate strategic triumph. As Colossians 2:15 declares with divine authority, "And having disarmed the powers and authorities, he made a public spectacle of them, triumphing over them by the cross." This was the decisive invasion, the D-Day of human history, the moment when the enemy's ultimate power was shattered, his chain of command irrevocably broken, and his ultimate defeat secured.

Therefore, you, soldier of Christ, fight *from* victory, not *for* victory. You do not wage war to earn God's favor or to secure salvation. You wage war because your salvation has already been secured, and you are called to enforce the victory Christ has already won in your life and in the world around you.

And if, as you journeyed through the chilling terrain of Phase IV, you saw your own reflection in its darkened corners – if you recognize the quiet prison, the abandoned weapons, the usurped throne of self, the life rebuilt in his image, or the comfortable, blinding darkness – hear this unyielding order from High Command: **It is not over.**

Even when a sheep, lost and wandering far from the flock, has given up all hope, the Shepherd has not. Jesus Himself poses the question in Matthew 18:12: "What do you think? If a man owns a hundred sheep, and one of them wanders away, will he not leave the ninety-nine on the hills and go to look for the one that wandered off?"

This is the very heart of your Commander. He is not a general who abandons His wounded. He is not a shepherd who writes off the one lost sheep, no matter how far it has strayed, no matter how entangled in briars it has become, no matter how much it smells of the world's filth. The Father's heart relentlessly pursues the one who is lost, and His grace is not merely sufficient; it is an inexhaustible arsenal of restoration. His arms are perpetually open. His forgiveness is boundless (1 John 1:9), ready to cleanse you from all unrighteousness. His power to resurrect dry bones (Ezekiel 37) is as mighty today as it ever was. You may feel broken, but you are not beyond redemption. You may feel disarmed, but the weapons of our warfare (2 Corinthians 10:4) are still available for your grasp. You may be blinded, but the Light of the World still shines.

Spiritual slavery is not your final destination unless *you choose it to be*. It is never too late to repent, to return, and to reclaim the ground that is rightfully His. The battle for your soul may have left you scarred and weary, but the opportunity for a glorious spiritual resurgence is always present. The first step, the decisive counter assault, is simply to **cry out to your Commander, to confess your compromise, and to take His outstretched hand**.

The battles, however, remain. This manual has equipped you to recognize the enemy's methods in Phases I, II, and III – his subtle reconnaissance, his insidious deceptions, his relentless attrition, and his efforts to isolate and suppress the will to resist. These are the daily engagements you will face. This is the ongoing mission: to enforce the victory Christ has already won in every sphere of your life. The war is won, but the mop-up operation, the enforcement of that victory, requires your vigilant, empowered participation.

Your final orders are clear:

Don your full armor. Each piece is a divine provision: the Belt of Truth to counter the enemy's lies; the Breastplate of Righteousness to guard your heart against shame; your Feet fitted with the Gospel of Peace, ready to advance His kingdom; the Shield of Faith to extinguish every fiery dart; the Helmet of Salvation to protect your mind; and the Sword of the Spirit, which is the Word of God, sharp and powerful, for both defense and offense (Ephesians 6:10-18).

Stay alert. Be sober-minded and watchful (1 Peter 5:8). The roaring lion seeks to devour, but he can only devour whom he finds unprepared.

Communicate constantly with High Command. Prayer is your direct line to the omnipotent Commander, your means of calling in air support, receiving intelligence, and strengthening your resolve. Pray in the Spirit on all occasions, with all kinds of prayers and requests (Ephesians 6:18).

Fight for unity. The Body of Christ is your impenetrable defensive perimeter, your ultimate support structure, and your combined arms force. Refuse to let the enemy sow division or isolate you from your fellow soldiers. Your unity is a direct assault on the kingdom of darkness.

Go and conquer. Live not in fear of the enemy, but in the power of the Holy Spirit who dwells within you (1 John 4:4). You are more than a conqueror through Christ Jesus (Romans 8:37). Enforce the enemy's defeat in your life, your family, your community, and your world, until the day the final trumpet sounds and our Commander returns to claim His eternal triumph.

The victory is **secured**.

Now, **fight**.

Appendix

Your Spiritual Armory
(Key Verses for Memorization)

THIS GUIDE HAS detailed the enemy's tactics across every phase of spiritual warfare. Knowing his schemes is crucial, but knowledge alone is not enough. You must be armed. Just as a soldier trains with their rifle, memorizing Scripture is the process of stockpiling your spiritual ammunition. These verses are the truth you will wield against the enemy's lies, the light you will shine into his darkness, and the promises you will stand upon when everything else feels uncertain. This is your personal armory. Load it into your mind, ready to deploy at a moment's notice.

I. The Commander's Authority & Overall Victory

Colossians 2:15: "And having disarmed the powers and authorities, he made a public spectacle of them, triumphing over them by the cross."

1 John 4:4: "You, dear children, are from God and have overcome them, because the one who is in you is greater than the one who is in the world."

Romans 8:37: "No, in all these things we are more than conquerors through him who loved us."

Romans 8:31: "What, then, shall we say in response to these things? If God is for us, who can be against us?"

John 10:10: "The thief comes only to steal and kill and destroy; I have come that they may have life, and have it to the full."

II. Truth, Discernment & The Word of God

Hebrews 4:12: "For the word of God is alive and active. Sharper than any double-edged sword, it penetrates even to dividing soul and spirit, joints and marrow; it judges the thoughts and attitudes of the heart."

John 8:31-32: "To the Jews who had believed him, Jesus said, 'If you hold to my teaching, you are really my disciples. Then you will know the truth, and the truth will set you free.'"

2 Timothy 3:16-17: "All Scripture is God-breathed and is useful for teaching, rebuking, correcting and training in righteousness, so that the servant of God may be thoroughly equipped for every good work."

1 John 4:1: "Dear friends, do not believe every spirit, but test the spirits to see whether they are from God, because many false prophets have gone out into the world." (Essential for discernment)

Proverbs 14:12: "There is a way that appears to be right, but in the end it leads to death." (Warning against deception)

III. Armor of God & Spiritual Strength

Ephesians 6:10-11: "Finally, be strong in the Lord and in his mighty power. Put on the full armor of God, so that you can take your stand against the devil's schemes."

Ephesians 6:13-18: "Therefore put on the full armor of God, so that when the day of evil comes, you may be able to stand your ground, and after you have done everything, to stand. Stand firm then, with the belt of truth buckled around your waist, with the breastplate of righteousness in place, and with your feet fitted with the readiness that comes from the gospel of peace. In addition to all this, take up the shield of faith, with which you can extinguish all the flaming arrows of the evil one. Take the helmet of salvation and the sword of the Spirit, which is the word of God. And pray in the Spirit on all occasions with all kinds of prayers and requests. With this in mind, be alert and always keep on praying for all the Lord's people."

2 Corinthians 10:3-4: "For though we live in the world, we do not wage war as the world does. The weapons we fight with are not the weapons of the world. On the contrary, they have divine power to demolish strongholds."

Psalm 27:1: "The LORD is my light and my salvation—whom shall I fear? The LORD is the stronghold of my life—of whom shall I be afraid?"

IV. Mind Warfare & Countering Deception

2 Corinthians 10:5: "We demolish arguments and every pretension that sets itself up against the knowledge of God, and we take captive every thought to make it obedient to Christ."

Philippians 4:8: "Finally, brothers and sisters, whatever is true, whatever is noble, whatever is right, whatever is pure, whatever is lovely, whatever is admirable—if anything is excellent or praiseworthy—think about such things."

John 8:44: "You belong to your father, the devil, and you want to carry out your father's desires. He was a murderer from the beginning, not holding to the truth, for there is no truth in him. When he lies, he speaks his native language, for he is a liar and the father of lies."

2 Corinthians 11:3: "But I am afraid that just as Eve was deceived by the serpent's cunning, your minds may somehow be led astray from your sincere and pure devotion to Christ."

2 Timothy 1:7: "For God has not given us a spirit of fear, but of power and of love and of a sound mind."

V. Overcoming Temptation & Sin

1 Corinthians 10:13: "No temptation has overtaken you except what is common to mankind. And God is faithful; he will not let you be tempted beyond what you can bear. But when you are tempted, he will also provide a way out so that you can endure it."

James 1:14: "but each person is tempted when they are dragged away by their own evil desire and enticed."

Romans 6:12-13: "Therefore do not let sin reign in your mortal body so that you obey its evil desires. Do not offer any part of yourself to sin as an instrument of wickedness, but rather offer yourselves to God as those who have been brought from death to life; and offer every part of yourself to him as an instrument of righteousness."

1 Corinthians 6:18: "Flee from sexual immorality. All other sins a person commits are outside the body, but whoever sins sexually sins against their own body."

1 John 1:9: "If we confess our sins, he is faithful and just and will forgive us our sins and purify us from all unrighteousness."

Psalm 119:11: "I have hidden your word in my heart that I might not sin against you."

VI. Identity in Christ & Freedom from Condemnation

Romans 8:1: "Therefore, there is now no condemnation for those who are in Christ Jesus."

2 Corinthians 5:17: "Therefore, if anyone is in Christ, the new creation has come: The old has gone, the new is here!"

Galatians 5:1: "It is for freedom that Christ has set us free. Stand firm, then, and do not let yourselves be burdened again by a yoke of slavery."

Romans 6:16: "Don't you know that when you offer yourselves to someone as obedient slaves, you are slaves of the one you obey—whether you are slaves to sin, which leads to death, or to obedience, which leads to righteousness?"

VII. Unity, Fellowship & Overcoming Isolation

Ecclesiastes 4:9-10: "Two are better than one, because they have a good return for their labor: If either of them falls down, one can help the other up. But pity anyone who falls and has no one to help them up!"

Galatians 6:2: "Carry each other's burdens, and in this way you will fulfill the law of Christ."

1 Corinthians 12:27: "Now you are the body of Christ, and each one of you is a part of it."

Hebrews 10:25: "Let us not give up meeting together, as some are in the habit of doing, but encouraging one another—and all the more as you see the Day approaching."

James 3:16: "For where you have envy and selfish ambition, there you find disorder and every evil practice."

Ephesians 4:27: "and do not give the devil a foothold."

VIII. Forgiveness & Reconciliation

Ephesians 4:32: "Be kind and compassionate to one another, forgiving each other, just as in Christ God forgave you."

Matthew 6:14-15: "For if you forgive other people when they sin against you, your heavenly Father will also forgive you. But if you do not forgive others their sins, your Father will not forgive your sins."

Hebrews 12:15: "See to it that no one falls short of the grace of God and that no bitter root grows up to cause trouble and defile many."

IX. God's Sovereignty & Faithfulness

Philippians 4:6-7: "Do not be anxious about anything, but in every situation, by prayer and petition, with thanksgiving, present your requests to God. And the peace of God, which transcends all understanding, will guard your hearts and your minds in Christ Jesus."

Isaiah 41:10: "So do not fear, for I am with you; do not be dismayed, for I am your God. I will strengthen you and help you; I will uphold you with my righteous right hand."

Jeremiah 29:11: "'For I know the plans I have for you,' declares the Lord, 'plans to prosper you and not to harm you, plans to give you hope and a future.'"

Matthew 18:12: "What do you think? If a man owns a hundred sheep, and one of them wanders away, will he not leave the ninety-nine on the hills and go to look for the one that wandered off?"

X. Readiness & Purpose

1 Peter 5:8: "Be alert and of sober mind. Your enemy the devil prowls around like a roaring lion looking for someone to devour."

Ephesians 2:1-2: "As for you, you were dead in your transgressions and sins, in which you used to live when you followed the ways of this world and of the ruler of the kingdom of the air, the spirit who is now at work in those who are disobedient."

Romans 12:2: "Do not conform to the pattern of this world, but be transformed by the renewing of your mind. Then you will be able to test and approve what God's will is—his good, pleasing and perfect will."

Galatians 6:9: "Let us not become weary in doing good, for at the proper time we will reap a harvest if we do not give up."

Luke 8:14: "The seed that fell among thorns stands for those who hear, but as they go on their way they are choked by life's worries, riches and pleasures, and they do not mature."

1 Corinthians 3:17: "If anyone destroys God's temple, God will destroy that person; for God's temple is sacred, and you together are that temple."

Endnotes

Phase I: The Pre-Conflict (Laying the Groundwork)

1. Clausewitz, Carl von. *On War*. Edited and translated by Michael Howard and Peter Paret. Princeton University Press, 1984, p. 69.

Chapter 1: The Reality of War – Endnotes

1. Sun Tzu, *The Art of War*, trans. Samuel B. Griffith (Oxford: Oxford University Press, 1963), 66–68.
2. Rick Atkinson, *Crusade: The Untold Story of the Persian Gulf War* (Boston: Houghton Mifflin, 1993), 89–94.
3. Michael R. Gordon and Bernard E. Trainor, *The Generals' War: The Inside Story of the Conflict in the Gulf* (Boston: Little, Brown, 1995), 291–296.
4. David Kahn, *Seizing the Enigma: The Race to Break the German U-Boats Codes, 1939–1943* (Boston: Houghton Mifflin, 1991), 252–257.
5. F.H. Hinsley and Alan Stripp, eds., *Codebreakers: The Inside Story of Bletchley Park* (Oxford: Oxford University Press, 1993), 34–39.
6. Ben Macintyre, *Double Cross: The True Story of the D-Day Spies* (New York: Crown, 2012), 112–119.

Chapter 2: Defining Objectives (Our Downfall) – Endnotes

1. Carl von Clausewitz, *On War*, ed. and trans. Michael Howard and Peter Paret (Princeton: Princeton University Press, 1976), 177–178.
2. Colin S. Gray, *Modern Strategy* (Oxford: Oxford University Press, 1999), 18–22.
3. Alistair Horne, *The Price of Glory: Verdun 1916* (New York: Penguin Books, 1993), 83–89.
4. Paul Jankowski, *Verdun: The Longest Battle of the Great War* (Oxford: Oxford University Press, 2014), 211–215.
5. Mark Clodfelter, *The Limits of Air Power: The American Bombing of North Vietnam* (Lincoln: University of Nebraska Press, 2006), 47–55.
6. Robert A. Pape, *Bombing to Win: Air Power and Coercion in War* (Ithaca: Cornell University Press, 1996), 174–178.
7. Michael B. Oren, *Six Days of War: June 1967 and the Making of the Modern Middle East* (New York: Oxford University Press, 2002), 163–170.
8. Chaim Herzog, *The Arab-Israeli Wars: War and Peace in the Middle East* (New York: Vintage Books, 1984), 214–220.
9. Saul David, *Operation Thunderbolt: Flight 139 and the Raid on Entebbe Airport* (New York: Little, Brown, 2015), 132–140.
10. Benny Morris, *Righteous Victims: A History of the Zionist-Arab Conflict, 1881–1999* (New York: Vintage, 2001), 392–394.
11. Hew Strachan, *The Direction of War: Contemporary Strategy in Historical Perspective* (Cambridge: Cambridge University Press, 2013), 101–104.

Chapter 3: The Art of Deception (Tactical Planning) – Endnotes

1. Edward N. Luttwak, *Strategy: The Logic of War and Peace* (Cambridge, MA: Belknap Press of Harvard University Press, 2001), 93–99.
2. Sun Tzu, *The Art of War*, trans. Samuel B. Griffith (Oxford: Oxford University Press, 1963).

3. Adrian Goldsworthy, *The Fall of Carthage: The Punic Wars 265–146 BC* (London: Cassell, 2000), 229–236.
4. Antony Beevor, *D-Day: The Battle for Normandy* (New York: Viking, 2009), 91–94.
5. Stephen E. Ambrose, *D-Day, June 6, 1944: The Climactic Battle of World War II* (New York: Simon & Schuster, 1994), 204–210.

Chapter 4: Coalition Building (The Enemy's Hierarchy) – Endnotes

1. Jeremy Black, *War and the World: Military Power and the Fate of Continents, 1450–2000*, Yale University Press, 1998.
2. Gerhard L. Weinberg, *A World at Arms: A Global History of World War II*, Cambridge University Press, 2005.
3. Williamson Murray and Allan R. Millett, *A War to Be Won: Fighting the Second World War, 1937–1945*, Belknap Press, 2000.
4. Rick Atkinson, *Crusade: The Untold Story of the Persian Gulf War*, Houghton Mifflin, 1993.
5. Clinton E. Arnold, *Powers of Darkness: Principalities & Powers in Paul's Letters* Downers Grove, IL: InterVarsity Press, 1992
6. Peter T. O'Brien, *The Letter to the Ephesians* (Grand Rapids: Eerdmans, 1999), 466–470; BDAG, ἀρχή (*archē* — "rule, authority, ruler"), 138–139
7. F. F. Bruce, *The Epistles to the Colossians, to Philemon, and to the Ephesians* (NICNT; Grand Rapids: Eerdmans, 1984), 406–407; BDAG, ἐξουσία (*exousia* — "authority, ruling power"), 353–355.
8. Andrew T. Lincoln, *Ephesians* (Word Biblical Commentary; Dallas: Word, 1990), 444–446; TDNT, κοσμοκράτωρ (*kosmokratōr* — "world-ruler, cosmic power"), 3:868–869.
9. John R. W. Stott, *God's New Society: The Message of Ephesians* (Downers Grove, IL: InterVarsity Press, 1979), 279–282; Louw-Nida, πνευματικός (*pneumatikos* — "pertaining to spirit, spiritual being"), 12.18.

Chapter 5: The Strategy of Deterrence (Preventing Conflict)

1. Lawrence Freedman, *Deterrence* (Cambridge: Polity Press, 2004), 33–37.
2. John J. Mearsheimer, *Conventional Deterrence* (Ithaca: Cornell University Press, 1983), 14–18.
3. Thomas C. Schelling, *Arms and Influence* (New Haven: Yale University Press, 1966), 235–241.
4. Lawrence Freedman, *The Evolution of Nuclear Strategy*, 3rd ed. (London: Palgrave Macmillan, 2003), 180–185.
5. John Lewis Gaddis, *The Cold War: A New History* (New York: Penguin Press, 2005), 52–56.
6. Adrian Goldsworthy, *How Rome Fell: Death of a Superpower* (New Haven: Yale University Press, 2009), 42–46.
7. Edward N. Luttwak, *The Grand Strategy of the Roman Empire: From the First Century A.D. to the Third* (Baltimore: Johns Hopkins University Press, 1976), 74–77.

Chapter 6: Preparation & Mobilization (Positioning His Forces)

1. Martin van Creveld, *Supplying War: Logistics from Wallenstein to Patton* (Cambridge: Cambridge University Press, 1977), 1–5.
2. John Andreas Olsen and Martin van Creveld, eds., *The Evolution of Operational Art: From Napoleon to the Present* (Oxford: Oxford University Press, 2011), 205–210.
3. Richard J. Evans, *The Third Reich in Power, 1933–1939* (New York: Penguin Press, 2005), 321–328.
4. William L. Shirer, *The Rise and Fall of the Third Reich* (New York: Simon & Schuster, 1960), 293–296.
5. Stephen E. Ambrose, *D-Day, June 6, 1944: The Climactic Battle of World War II* (New York: Simon & Schuster, 1994), 105–112.
6. Antony Beevor, *D-Day: The Battle for Normandy* (New York: Viking, 2009), 25–29.
7. Max Hastings, *Overlord: D-Day and the Battle for Normandy* (New York: Simon & Schuster, 1984), 50–54.

Phase II: The Initial Phase (The Assault)

1. Carl von Clausewitz, *On War* (Princeton: Princeton University Press, 1984), Book II, Ch. 2, pp. 117–119.
2. Harlan K. Ullman and James P. Wade, *Shock and Awe: Achieving Rapid Dominance* (Washington, D.C.: National Defense University, 1996), pp. xiii–xv.
3. Giulio Douhet, *The Command of the Air* (Washington, D.C.: Office of Air Force History, 1983), Part I, Ch. 2, pp. 25–28.
4. U.S. Department of the Army, *Psychological Operations Tactics, Techniques, and Procedures* (FM 3-05.301, Washington, D.C.: Headquarters, Department of the Army, 2007), pp. 1-3–1-4.

Chapter 7: Gaining Air Superiority (Controlling the Narrative)

1. John Andreas Olsen, *Airpower Reborn: The Strategic Concepts of John Warden and John Boyd* (Annapolis: Naval Institute Press, 2015), 15–20.
2. Chaim Herzog, *The Arab-Israeli Wars: War and Peace in the Middle East* (New York: Vintage, 1984), 219–225.
3. Michael B. Oren, *Six Days of War: June 1967 and the Making of the Modern Middle East* (Oxford: Oxford University Press, 2002), 171–178.
4. Richard Overy, *The Battle of Britain: Myth and Reality* (London: Penguin, 2000), 56–61.
5. Stephen Bungay, *The Most Dangerous Enemy: A History of the Battle of Britain* (London: Aurum Press, 2000), 143–150.
6. Williamson Murray, *Luftwaffe: Strategy for Defeat, 1933–1945* (Maxwell Air Force Base: Air University Press, 1983), 90–94.

Chapter 8: Information Warfare (Propaganda & Psychological Operations)

1. U.S. Department of Defense, *Joint Publication 3-13: Information Operations* (Washington, DC: Joint Chiefs of Staff, 2012), I-1–I-3.

2. Philip M. Taylor, *Munitions of the Mind: A History of Propaganda from the Ancient World to the Present Era* (Manchester: Manchester University Press, 2003), 5–8.
3. Garth S. Jowett and Victoria O'Donnell, *Propaganda & Persuasion*, 7th ed. (Thousand Oaks: SAGE Publications, 2019), 6–12.
4. Richard Lucas, *Axis Sally: The American Voice of Nazi Germany* (Havertown, PA: Casemate, 2010), 101–107.
5. Stanley Mead, *Tokyo Rose: Orphan of the Pacific* (New York: Charles Scribner's Sons, 1971), 87–92.
6. Paul M.A. Linebarger, *Psychological Warfare* (Washington, D.C.: Infantry Journal Press, 1948), 175–178.
7. Arch Puddington, *Broadcasting Freedom: The Cold War Triumph of Radio Free Europe and Radio Liberty* (Lexington: University Press of Kentucky, 2000), 44–50.
8. A. Ross Johnson, *Radio Free Europe and Radio Liberty: The CIA Years and Beyond* (Stanford: Stanford University Press, 2010), 233–240.

Chapter 9: Suppression of Enemy Defenses (Silencing the Truth)

1. Benjamin S. Lambeth, *The Transformation of American Air Power* (Cornell University Press, 2000), p. 187.
2. Bill Sweetman, *Lockheed F-117 Stealth Fighter* (Motorbooks International, 1989), pp. 45–46.
3. Steve Davies, *USAF Wild Weasels: SAM Suppression Tactics 1965–SAM Desert Storm* (Osprey Publishing, 2005), pp. 22–25.
4. Rick Atkinson, *Crusade: The Untold Story of the Persian Gulf War* (Houghton Mifflin, 1993), pp. 73–74.
5. Norman Friedman, *Desert Victory: The War for Kuwait* (Naval Institute Press, 1991), p. 124.
6. Richard P. Hallion, *Storm Over Iraq: Air Power and the Gulf War* (Smithsonian Institution Press, 1992), pp. 105–107.
7. Zeev Schiff, *A History of the Israeli Army: 1874 to the Present* (Macmillan, 1985), pp. 312–314.

Chapter 10: The Initial Offensive (Exploiting Our Weaknesses)

1. Harlan K. Ullman & James P. Wade, *Shock and Awe: Achieving Rapid Dominance* (National Defense University Press, 1996), pp. xiii–xiv.
2. Julian Jackson, *The Fall of France: The Nazi Invasion of 1940* (Oxford University Press, 2003), pp. 87–90.
3. Karl-Heinz Frieser, *The Blitzkrieg Legend: The 1940 Campaign in the West* (Naval Institute Press, 2005), pp. 104–106.
4. Alistair Horne, *To Lose a Battle: France 1940* (Penguin, 1969), pp. 327–330.
5. Ernest R. May, *Strange Victory: Hitler's Conquest of France* (Hill and Wang, 2000), pp. 289–293.
6. Roger A. Beaumont, *Military Elites: Special Fighting Units in the Modern World* (Stein and Day, 1974), pp. 63–64.
7. James Lucas, *Storming Eagles: German Airborne Forces in World War Two* (Cassell, 2003), pp. 37–40.
8. Clay Blair, *Hitler's U-Boat War: The Hunters, 1939–1942* (Random House, 1996), p. 95.

Chapter 11: Nava Operations – Sea Control (Infiltrating Our Hearts)

1. Alfred Thayer Mahan, *The Influence of Sea Power upon History, 1660–1783* (Little, Brown, 1890), pp. 25–28.
2. John Gooch, *Naval Blockades and Seapower: Strategies and Counter-Strategies, 1805–2005* (Routledge, 2006), pp. 3–5.
3. Andrew Lambert, *Nelson: Britannia's God of War* (Faber & Faber, 2004), pp. 269–274.
4. N.A.M. Rodger, *The Command of the Ocean: A Naval History of Britain, 1649–1815* (W.W. Norton, 2005), pp. 544–547.
5. Craig L. Symonds, *The Civil War at Sea* (Oxford University Press, 2012), pp. 18–22.
6. James M. McPherson, *Battle Cry of Freedom: The Civil War Era* (Oxford University Press, 1988), pp. 351–353.
7. David G. Surdam, *Northern Naval Superiority and the Economics of the American Civil War* (University of South Carolina Press, 2001), pp. 62–65.

Phase III: The Sege

1. Christopher Duffy, *Siege Warfare: The Fortress in the Early Modern World 1494–1660* (London: Routledge, 1996), pp. 1–3.
2. John Keegan, *The Face of Battle* (New York: Viking Press, 1976), pp. 130–132.
3. Richard Holmes, *Acts of War: The Behavior of Men in Battle* (New York: Free Press, 1985), pp. 212–214.
4. Martin van Creveld, *Supplying War: Logistics from Wallenstein to Patton* (Cambridge: Cambridge University Press, 2004), pp. 27–29.

Chapter 12: Degrading Enemy Capabilities (Severing the Connection)

1. Milan N. Vego, *Joint Operational Warfare: Theory and Practice* (Naval War College, 2007), pp. X-13 – X-16.
2. John Arquilla and David Ronfeldt, *Cyberwar is Coming!* (RAND, 1993), pp. 29–34.
3. Stephen L. McFarland and Wesley Phillips Newton, *To Command the Sky: The Battle for Air Superiority over Germany, 1942–1944* (Smithsonian Institution Press, 1991), pp. 305–309.
4. Matthew Cobb, *The Resistance: The French Fight Against the Nazis* (Simon & Schuster, 2009), pp. 187–193.
5. Max Hastings, *Overlord: D-Day and the Battle for Normandy* (Vintage, 1984), pp. 163–169.
6. Julius Caesar, *The Gallic War*, trans. Carolyn Hammond (Oxford World's Classics, 1996), Book VII, §§68–72.
7. Adrian Goldsworthy, *Caesar: Life of a Colossus* (Yale University Press, 2006), pp. 356–362.

Chapter 13: Destroying Key Infrastructure (The Foundations of Faith)

1. Milan N. Vego, *Joint Operational Warfare: Theory and Practice* (Naval War College, 2007), pp. VIII-7 – VIII-10.
2. Richard Overy, *The Bombing War: Europe 1939–1945* (Penguin, 2013), pp. 398–404.
3. Williamson Murray, *Strategy for Defeat: The Luftwaffe 1933–1945* (Air University Press, 1983), pp. 165–172.

4. Mark Grimsley, *The Hard Hand of War: Union Military Policy toward Southern Civilians, 1861–1865* (Cambridge University Press, 1995), pp. 193–198.
5. James M. McPherson, *Battle Cry of Freedom: The Civil War Era* (Oxford University Press, 1988), pp. 720–726.

Chapter 14: Attrition Warfare (Wearing Us Down)

1. Hew Strachan, *The First World War* (Penguin, 2004), pp. 191–200.
2. Antony Beevor, *Stalingrad* (Penguin, 1998), pp. 147–155.
3. Clay Blair, *Hitler's U-Boat War: The Hunters, 1939–1942* (Random House, 1996), pp. 89–94.
4. Ibid.
5. Marc Milner, *Battle of the Atlantic* (University of Kansas Press, 2003), pp. 215–220.
6. Lewis Sorley, *Westmoreland: The General Who Lost Vietnam* (Houghton Mifflin, 2011), pp. 217–225.
7. George C. Herring, *America's Longest War: The United States and Vietnam, 1950–1975* (McGraw Hill, 2002), pp. 224–232.

Chapter 15: Securing Key Terrain (Claiming Territory in Ourt Lives)

1. Kaplan, Robert D. *The Revenge of Geography: What the Map Tells Us About Coming Conflicts and the Battle Against Fate* (Random House, 2012).
2. Carl von Clausewitz, *On War*, ed. and trans. Michael Howard and Peter Paret (Princeton: Princeton University Press, 1976), Book 3, Chapter 8.
3. Ibid.
4. James R. Arnold, *The First Domino: The Battle of Hamburger Hill, May 1969* (New York: William Morrow, 1991), 142–167.
5. Samuel Zaffiri, *Hamburger Hill: May 11–20, 1969* (Novato, CA: Presidio Press, 1988), 223–240.
6. U.S. Army Center of Military History, *Combat Operations: Stemming the Tide, May 1965 to October 1966* (Washington, D.C.: Government Printing Office, 1991).
7. Coddington, Edwin B. *The Gettysburg Campaign: A Study in Command.* Charles Scribner's Sons, 1968.

Chapter 16: Isolating the Enemy (Isolation Us from the Body of Christ)

1. Department of the Navy. *MCDP 1, Warfighting.* Washington, DC: Headquarters, U.S. Marine Corps, 1997. (See sections on "Maneuver Warfare," "Exploiting Weakness," and "Center of Gravity").
2. Robert M. Citino, *The German Way of War: From the Thirty Years' War to the Third Reich* (Lawrence: University Press of Kansas, 2005), 276–282.
3. Adrian Goldsworthy, *The Punic Wars* (London: Cassell, 2000), 297–315.
4. Antony Beevor, *Stalingrad: The Fateful Siege, 1942–1943* (New York: Viking, 1998), 263–318.

Chapter 17: Counterinsurgency (Suppressing Our Will to Resist)

1. U.S. Army and Marine Corps, *FM 3-24: Counterinsurgency* (Washington, DC: Department of the Army, 2006), 1–3.
2. David Kilcullen, *Counterinsurgency* (Oxford: Oxford University Press, 2010), 1–12.
3. U.S. Army & Marine Corps, *Counterinsurgency Field Manual: FM 3-24* (Chicago: University of Chicago Press, 2007), xxv–xxviii.
4. U.S. Army and Marine Corps, *FM 3-24: Counterinsurgency* (Washington, DC: Department of the Army, 2006).
5. Ibid.
6. Martin Goodman, *Rome and Jerusalem: The Clash of Ancient Civilizations* (London: Penguin, 2007), 385–402.
7. Peter Schäfer, *The Bar Kokhba War Reconsidered: New Perspectives on the Second Jewish Revolt Against Rome* (Tübingen: Mohr Siebeck, 2003), 9–27.
8. Richard Stubbs, *Hearts and Minds in Guerrilla Warfare: The Malayan Emergency, 1948–1960* (Singapore: Oxford University Press, 1989), 170–192.
9. Kumar Ramakrishna, *Emergency Propaganda: The Winning of Malayan Hearts and Minds, 1948–1958* (Richmond, Surrey: Curzon Press, 2002), 88–105.

Chapter 18: Establishing Security (Controlling Our Thoughts and Actions)

1. Martin Hengel, *Crucifixion: In the Ancient World and the Folly of the Message of the Cross* (Philadelphia: Fortress Press, 1977), 22–32.
2. John O. Koehler, *Stasi: The Untold Story of the East German Secret Police* (Boulder, CO: Westview Press, 1999), 35–56.
3. Jens Gieseke, *The History of the Stasi: East Germany's Secret Police, 1945–1990* (New York: Berghahn Books, 2014), 78.
4. Hubertus Knabe, *Die Täter sind unter uns: Über das Schönreden der SED-Diktatur* (Berlin: Propyläen, 2007), 112–129.
5. Martin E. P. Seligman, *Helplessness: On Depression, Development, and Death* (San Francisco: W. H. Freeman, 1975), 28–45.

Chapter 19: Disarmament and Demobilization (Stripping Us of Our Spiritual Weapons)

1. United Nations, *Disarmament, Demobilization and Reintegration of Ex-Combatants in Peacekeeping Operations: Principles and Guidelines* (New York: UN DPKO, 2000), 5–9.
2. James Dobbins, Seth G. Jones, Keith Crane, et al., *The UN's Role in Nation-Building: From the Congo to Iraq* (Santa Monica, CA: RAND Corporation, 2005), 73–78.
3. Ibid.
4. John W. Dower, *Embracing Defeat: Japan in the Wake of World War II* (New York: W.W. Norton & Company, 1999), 65–88.
5. Richard B. Finn, *Winners in Peace: MacArthur, Yoshida, and Postwar Japan* (Berkeley: University of California Press, 1992), 33–47.
6. Kenneth L. Port, *Article 9 of the Japanese Constitution: A Documentary History of Its Past and Present* (Armonk, NY: M.E. Sharpe, 1998), 14–23.

Chapter 20: Establishing a New Government (Replacing God's Authority with His Own)

1. Brian Taylor, *The Soviet Puppet States of Eastern Europe* (New Haven: Yale University Press, 2010), 41–46.
2. Hannah Arendt, *The Origins of Totalitarianism* (New York: Harcourt, Brace, 1951), 211–215.
3. Robert O. Paxton, *Vichy France: Old Guard and New Order, 1940–1944* (New York: Columbia University Press, 2001), 3–27.
4. Julian Jackson, *France: The Dark Years, 1940–1944* (Oxford: Oxford University Press, 2001), 126–147.
5. Michael R. Marrus and Robert O. Paxton, *Vichy France and the Jews* (Stanford: Stanford University Press, 1995), 61–84.

Chapter 21: Reconstruction and Development (Rebuilding Our Lives in Satan's Image)

1. James Dobbins, John G. McGinn, Keith Crane, et al., *America's Role in Nation-Building: From Germany to Iraq* (Santa Monica, CA: RAND Corporation, 2003), 135–142.
2. U.S. Army, *FM 3-07: Stability Operations* (Washington, DC: Department of the Army, 2008), 1-4–1-7.
3. G. John Ikenberry, *After Victory: Institutions, Strategic Restraint, and the Rebuilding of Order After Major Wars* (Princeton: Princeton University Press, 2001), 183–189.
4. Michael J. Hogan, *The Marshall Plan: America, Britain, and the Reconstruction of Western Europe, 1947–1952* (Cambridge: Cambridge University Press, 1987), 24–45.
5. John W. Dower, *Embracing Defeat: Japan in the Wake of World War II* (New York: W.W. Norton & Company, 1999), 65–98.
6. Takemae Eiji, *Inside GHQ: The Allied Occupation of Japan and Its Legacy*, trans. Robert Ricketts and Sebastian Swann (New York: Continuum, 2002), 225–248.

Chapter 22: Winning Hearts and Minds (Blinding Us to the Truth)

1. U.S. Army and Marine Corps, *FM 3-24: Counterinsurgency* (Washington, DC: Department of the Army, 2006), 89–94.
2. David Galula, *Counterinsurgency Warfare: Theory and Practice* (Westport, CT: Praeger, 1964), 74–82.
3. Nicholas J. Cull, *The Cold War and the United States Information Agency: American Propaganda and Public Diplomacy, 1945–1989* (Cambridge: Cambridge University Press, 2008), 92–115.
4. Frances Stonor Saunders, *The Cultural Cold War: The CIA and the World of Arts and Letters* (New York: The New Press, 1999), 141–176.
5. Vladislav M. Zubok, *A Failed Empire: The Soviet Union in the Cold War from Stalin to Gorbachev* (Chapel Hill: University of North Carolina Press, 2007), 55–83.

Bibliography

Ambrose, Stephen E. *D-Day, June 6, 1944: The Climactic Battle of World War II*. New York: Simon & Schuster, 1994.

Ambrose, Stephen E. *D-Day: The Battle for Normandy*. New York: Viking, 2009.

Arendt, Hannah. *The Origins of Totalitarianism*. New York: Harcourt, Brace, 1951.

Arquilla, John, and David Ronfeldt. *Cyberwar is Coming!* Santa Monica, CA: RAND Corporation, 1993.

Arquilla, John, and David Ronfeldt. *The Emergence of Noopolitik: Toward an American Information Strategy*. Santa Monica, CA: RAND Corporation, 1999.

Atkinson, Rick. *Crusade: The Untold Story of the Persian Gulf War*. Boston: Houghton Mifflin, 1993.

Beevor, Antony. *D-Day: The Battle for Normandy*. New York: Viking, 2009.

Beevor, Antony. *Stalingrad*. London: Penguin, 1998.

Black, Jeremy. *War and the World: Military Power and the Fate of Continents, 1450–2000*. New Haven: Yale University Press, 1998.

Blair, Clay. *Hitler's U-Boat War: The Hunters, 1939–1942*. New York: Random House, 1996.

Boot, Max. *Invisible Armies: An Epic History of Guerrilla Warfare from Ancient Times to the Present*. New York: W.W. Norton, 2013.

Bungay, Stephen. *The Most Dangerous Enemy: A History of the Battle of Britain*. London: Aurum Press, 2000.

Clausewitz, Carl von. *On War*. Edited and translated by Michael Howard and Peter Paret. Princeton: Princeton University Press, 1976.

Colletta, Nat J., and Robert Muggah. *Light Weapons and Civil Conflict: Controlling the Tools of Violence*. London: Routledge, 2009.

Creveld, Martin van. *The Transformation of War*. New York: Free Press, 1991.

Cull, Nicholas J. *The Cold War and the United States Information Agency: American Propaganda and Public Diplomacy, 1945–1989*. Cambridge: Cambridge University Press, 2008.

Department of the Army. *FM 3-07: Stability Operations*. Washington, DC: Headquarters, Department of the Army, 2008.

Department of the Army. *FM 3-24: Counterinsurgency*. Washington, DC: Headquarters, Department of the Army, 2006.

Department of the Army. *FM 3-05.301: Psychological Operations Process Tactics, Techniques, and Procedures*. Washington, DC: Headquarters, Department of the Army, 2007.

Department of the Navy. *MCDP 1: Warfighting*. Washington, DC: Headquarters, U.S. Marine Corps, 1997.

Dobbins, James, John G. McGinn, Keith Crane, et al. *America's Role in Nation-Building: From Germany to Iraq*. Santa Monica, CA: RAND Corporation, 2003.

Dobbins, James, Seth G. Jones, Keith Crane, and Beth Cole

DeGrasse. *The Beginner's Guide to Nation-Building*. Santa Monica, CA: RAND Corporation, 2007.

Dower, John W. *Embracing Defeat: Japan in the Wake of World War II*. New York: W.W. Norton, 1999.

Duffy, Christopher. *Siege Warfare: The Fortress in the Early Modern World 1494–1660*. London: Routledge, 1996.

Freedman, Lawrence. *Deterrence*. Cambridge: Polity Press, 2004.

Freedman, Lawrence. *The Evolution of Nuclear Strategy*. 3rd ed. London: Palgrave Macmillan, 2003.

Galula, David. *Counterinsurgency Warfare: Theory and Practice*. Westport, CT: Praeger Security International, 1964.

Goldsworthy, Adrian. *Caesar: Life of a Colossus*. New Haven: Yale University Press, 2006.

Goldsworthy, Adrian. *The Fall of Carthage: The Punic Wars 265–146 BC*. London: Cassell, 2000.

Grimsley, Mark. *The Hard Hand of War: Union Military Policy toward Southern Civilians, 1861–1865*. Cambridge: Cambridge University Press, 1995.

Hastings, Max. *Overlord: D-Day and the Battle for Normandy*. New York: Simon & Schuster, 1984.

Herring, George C. *America's Longest War: The United States and Vietnam, 1950–1975*. New York: McGraw Hill, 2002.

Herzog, Chaim. *The Arab-Israeli Wars: War and Peace in the Middle East*. New York: Vintage Books, 1984.

Hoffman, Frank G. *Conflict in the 21st Century: The Rise of Hybrid Wars*. Arlington, VA: Potomac Institute for Policy Studies, 2007.

Ikenberry, G. John. *After Victory: Institutions, Strategic Restraint, and the Rebuilding of Order After Major Wars*. Princeton: Princeton University Press, 2001.

Johnson, A. Ross. *Radio Free Europe and Radio Liberty: The CIA Years and Beyond.* Stanford: Stanford University Press, 2010.

Jowett, Garth S., and Victoria O'Donnell. *Propaganda & Persuasion.* 7th ed. Thousand Oaks: SAGE Publications, 2019.

Kilcullen, David. *Counterinsurgency.* Oxford: Oxford University Press, 2010.

Kilcullen, David. *The Accidental Guerrilla: Fighting Small Wars in the Midst of a Big One.* Oxford: Oxford University Press, 2009.

Linebarger, Paul M.A. *Psychological Warfare.* Washington, D.C.: Infantry Journal Press, 1948.

Luttwak, Edward N. *Strategy: The Logic of War and Peace.* Cambridge, MA: Belknap Press of Harvard University Press, 2001.

Mazower, Mark. *Hitler's Empire: Nazi Rule in Occupied Europe.* New York: Penguin Press, 2008.

McPherson, James M. *Battle Cry of Freedom: The Civil War Era.* New York: Oxford University Press, 1988.

Mumford, Andrew. *The Counter-Insurgency Myth: The British Experience of Irregular Warfare.* London: Routledge, 2012.

Murray, Williamson, and Mark Grimsley, eds. *The Making of Strategy: Rulers, States, and War.* Cambridge: Cambridge University Press, 1994.

Nagl, John A. *Learning to Eat Soup with a Knife: Counterinsurgency Lessons from Malaya and Vietnam.* Chicago: University of Chicago Press, 2005.

Oren, Michael B. *Six Days of War: June 1967 and the Making of the Modern Middle East.* Oxford: Oxford University Press, 2002.

Overy, Richard. *The Battle of Britain: Myth and Reality.* London: Penguin, 2000.

Overy, Richard. *The Bombing War: Europe 1939–1945*. London: Penguin, 2013.

Overy, Richard. *Why the Allies Won*. New York: W.W. Norton, 1996.

Pape, Robert A. *Bombing to Win: Air Power and Coercion in War*. Ithaca: Cornell University Press, 1996.

Saunders, Frances Stonor. *The Cultural Cold War: The CIA and the World of Arts and Letters*. New York: The New Press, 1999.

Smith, Rupert. *The Utility of Force: The Art of War in the Modern World*. New York: Knopf, 2007.

Stiglitz, Joseph E. *Globalization and Its Discontents*. New York: W.W. Norton, 2002.

Strachan, Hew. *The First World War*. London: Penguin, 2004.

Sun Tzu. *The Art of War*. Translated by Samuel B. Griffith. Oxford: Oxford University Press, 1963.

Taylor, Brian. *The Soviet Puppet States of Eastern Europe*. New Haven: Yale University Press, 2010.

Thompson, Sir Robert. *Defeating Communist Insurgency: Experiences from Malaya and Vietnam*. New York: Praeger, 1966.

Ullman, Harlan K., and James P. Wade. *Shock and Awe: Achieving Rapid Dominance*. Washington, D.C.: National Defense University, 1996.

United Nations Department of Peacekeeping Operations. *Disarmament, Demobilization and Reintegration of Ex-Combatants in Peacekeeping Operations: Principles and Guidelines*. New York: United Nations, 2000.

U.S. Department of Defense. *Joint Publication 3-13: Information Operations*. Washington, DC: Joint Chiefs of Staff, 2012.

Glossary of Terms

Accusation — In law/intelligence, a formal charge; in spiritual warfare, Satan's prosecutorial tactic to condemn and immobilize the believer (Rev 12:10), often leveraging real failures to block repentance and joy.

Air Superiority — Dominant control of the air that enables freedom of action; spiritually, dominance of the **narrative space** (media, thought-life) so truth is muted and lies circulate unchecked.

Ambush — Surprise attack from concealment; spiritually, sudden temptation arranged at predictable weak moments or places.

Annihilation Strategy — Aim to destroy an enemy's force outright; spiritually, campaigns to erase faith, identity, or hope rather than merely hinder.

Apathy (Spiritual Numbness) — Loss of feeling/concern; spiritually induced "chemical warfare" that desensitizes conscience and stalls repentance.

Armor of God — God's issued kit (Eph 6:10–18); spiritually, truth, righteousness, readiness, faith, salvation, and Scripture employed actively, not decoratively.

Attrition Warfare — Wearing an enemy down through sustained losses; spiritually, relentless, low-grade pressures that exhaust will and discipline over time.

Belt of Truth — Load-bearing gear that stabilizes the soldier;

spiritually, truth that holds together discernment and integrity.

Betrayal Operations — Turning insiders; spiritually, exploiting trusted voices (friends, teachers, leaders) to spread lies or normalize sin.

Blockade — Cutting off movement and supplies by sea/land; spiritually, choking prayer, Word intake, and fellowship to starve the soul.

Body Count Metrics — Measuring success by casualties; spiritually, hollow "wins" (outward activity, numbers) that ignore transformation.

Breach — Opening in a defense; spiritually, unresolved sin, bitterness, or pride that gives access (Eph 4:27).

Campaign — Linked operations toward a strategic end; spiritually, sequenced temptations, lies, and pressures serving a larger objective.

Center of Gravity (COG) — The source of a force's power; spiritually, the Gospel and our union with Christ—primary targets for deception and doubt.

Checkpoints — Security control points; spiritually, mental "no-go" zones where shame forbids thinking about calling, holiness, or confession.

Chokepoint — Narrow terrain that controls movement; spiritually, key habits/moments (late night, loneliness, pain) where decisions funnel.

Coalition (Enemy) — Allies coordinating toward a shared goal; spiritually, an organized hierarchy of demonic powers (Eph 6:12) and cultural systems.

Command and Control (C2) — Direction of forces; spiritually, God's authority and guidance accessed by prayer/Word—disrupted by sin or pride.

Counterinsurgency (COIN) — Suppressing rebellion via force and persuasion; spiritually, tactics to crush a believer's will to resist and make compromise feel permanent.

Counterintelligence (CI) — Detecting/degrading enemy intel; spiritually, proactive discernment, confession, and accountability that exposes lies and plots.

Cover and Concealment — Protection from fire/observation; spiritually, wise boundaries and discretion vs. hiding sin (which empowers it).

Decapitation Strike — Removing leadership; spiritually, discrediting Christ's authority or sabotaging leaders to scatter the flock.

Deception Operations — Misleading to gain advantage; spiritually, half-truths and misused Scripture that distort God's character and will.

Deep Battle — Striking beyond the front to disrupt reserves; spiritually, attacking identity, purpose, and hope, not just behavior.

Demobilization — Disbanding forces/roles; spiritually, convincing believers to retire from service and witness.

Denial (Area/Access) — Preventing use of an asset; spiritually, choking access to prayer, Scripture, or community.

Deterrence — Preventing action through fear/cost; spiritually, pre-emptive cynicism, mockery, or "it won't work" narratives that keep people from God.

Disarmament — Confiscating weapons; spiritually, turning Scripture, prayer, faith, and fellowship into neglected, "unused" gear.

Disinformation — False information seeded as truth; spiritually, cultural maxims that sound wise but oppose Scripture.

Doctrine (Military) — Agreed principles of war; spiritually, sound teaching that guides faithful practice—and a prime target for dilution.

Encirclement (Kesselschlacht) — Surrounding to cut off and crush; spiritually, isolation from the Body and resources leading to despair.

Enfilade Fire — Fire along the long axis of a target; spiritually, targeted lies that sweep across multiple life areas (identity, purpose, relationships).

Fire Superiority — Dominant, sustained firepower; spiritually, a flood of narratives and temptations that keep conscience pinned down.

Foothold — Small seized position for expansion; spiritually, tolerated sin, unforgiveness, or a lie we agree with (Eph 4:27).

Forward Operating Base (FOB) — Secured site enabling ops; spiritually, a stronghold within the heart from which attacks are launched.

Garrison — Troops stationed to hold ground; spiritually, entrenched lies or habits occupying mind and emotions.

Gospel-Plus — Adding requirements to grace; spiritually, legalism/performancism that drains joy and assurance (Eph 2:8–9).

Guerrilla Warfare — Small, irregular attacks; spiritually, sporadic temptations and micro-compromises that cumulatively derail devotion.

Hearts and Minds — Winning popular allegiance; spiritually, seducing affections and thinking so captivity feels like flourishing (2 Cor 4:4).

Idolatry — Worship of created things; spiritually, good gifts (career, family, cause) made ultimate, replacing God at the center.

Information Operations (IO) — Coordinated use of info; spiritually, synchronized lies, distractions, and cultural scripts shaping belief and behavior.

Insurgency — Armed resistance to rule; spiritually, the Holy Spirit-empowered will resisting sin and reclaiming ground in Christ.

Isolation — Cutting off from support; spiritually, withdrawing from church, mentors, and accountability—an enemy priority.

Learned Helplessness — Conditioned inaction after failure; spiritually, "I can't change" despair that refuses grace and effort.

Lines of Communication (LOCs) — Routes for supply/command; spiritually, prayer, Word, sacraments, fellowship—lifelines to be protected.

Logistics — Moving and sustaining forces; spiritually, rhythms and disciplines that nourish faith and endurance.

Maskirovka — Russian doctrine of deception/denial; spiritually, layered misdirection that hides sin and blurs truth.

Maneuver Warfare — Out-positioning to collapse will; spiritually, agile obedience that sidesteps traps and shifts to God's advantage.

Milice (Vichy) — Collaborationist police; spiritually, internalized self-condemnation policing the soul on the enemy's behalf.

Morale — Confidence/will to fight; spiritually, hope in Christ's promises—courage the enemy seeks to erode.

Objective (End State) — Concrete aim of an operation; spiritually, conformity to Christ vs. the enemy's aim: disbelief, despair, disunity.

Operational Art — Linking tactics to strategy; spiritually, integrating daily practices to serve lifelong holiness and mission.

Order of Battle (ORBAT) — Identification of enemy units; spiritually, awareness of personal vulnerabilities and the enemy's patterns.

Pacification — Making a population accept control; spiritually, converting a struggling believer into a contented captive.

Puppet Regime — Local government controlled by occupier; spiritually, Self installed on the heart's throne while the enemy rules from the shadows.

Psychological Operations (PSYOP) — Shaping perceptions/behavior; spiritually, the enemy's whispers of shame, fear, inevitability, and pride.

Reconnaissance (ISR) — Intelligence collection; spiritually, the enemy's long study of habits, wounds, and triggers to tailor attacks.

Red Teaming — Thinking like the adversary; spiritually, Scriptural self-examination to identify likely approaches and shore up defenses.

Repentance — Change of mind/turning; spiritually, decisive re-alignment with God that breaks legal ground for the enemy.

Resupply — Restoration of ammo/fuel; spiritually, Sabbath, prayer, Word, and fellowship that replenish grace and grit.

Rules of Engagement (ROE) — Limits on force use; spiritually, biblical boundaries guiding speech, desires, and actions.

Sanctification — Ongoing growth in holiness; spiritually, Spirit-empowered transformation the enemy resists through drift, distraction, and doubt.

SEAD (Suppression of Enemy Air Defenses) — Neutralizing air defenses; spiritually, silencing truth-tellers and discouraging prayer to clear the air for lies.

Self-Condemnation — Adopting the accuser's voice; spiritually, partnering with shame instead of receiving grace (Rom 8:1).

Shock and Awe (Rapid Dominance) — Overwhelming force to paralyze response; spiritually, blitz attacks meant to stampede into rash compromise.

Siege — Surrounding to starve and breach; spiritually, long pressure that isolates, exhausts, and normalizes defeat.

Situational Awareness (SA) — Real-time understanding of environment; spiritually, watchfulness over thoughts, influences, and motives.

Special Operations Forces (SOF) — Small, elite units.

Spiritual Disciplines — Rhythms that cultivate grace (prayer, Scripture, worship, fellowship, service, fasting); spiritually, your supply chain.

Spiritual Dryness — Felt absence of consolation; spiritually, a testing ground to walk by faith, not by sight.

Stronghold — Fortified position; spiritually, entrenched lies, sins, or patterns defended by arguments and pretensions (2 Cor 10:4–5).

Suppression — Reducing effective response; spiritually, numbing conscience and muting witness.

Targeting Cycle — Find, fix, finish, exploit, analyze; spiritually, how the enemy identifies and revisits weak points unless patterns change.

Theater (Battlespace) — Area of operations; spiritually, the whole of life—mind, body, relationships, vocation, church, culture.

Unity of Command/Effort — Aligned leadership/actions; spiritually, Christ's headship and the church's coordinated love and mission.

Whitewashed Tomb — Beautiful exterior, dead interior (Matt 23:27); spiritually, a polished but hollow life rebuilt on sand.

Zersetzung — Stasi method of psychological "decomposition"; spiritually, subtle manipulations that unravel sanity, relationships, and purpose.

About the Author

SHANE W. CUNNINGHAM is uniquely equipped to bridge the disciplines of military strategy and spiritual warfare. A United States Marine, Shane served in intelligence and counterintelligence roles at the highest levels of national security, including assignments with the Pentagon and specialized agencies. His career demanded an intimate understanding of strategy, deception, and unseen operations—skills that would later shape his perspective on the spiritual battlefield.

Following years of service in the military and government, Shane experienced a profound call to full-time Christian ministry. Together with his wife, Jessica, they left behind high-profile careers to dedicate their lives to discipleship, youth ministry, and international missions. Today, Shane leads a nonprofit based in Texas, that expands and supports orphanages across Indonesia while also shepherding local churches.

As a husband and father, Shane's ministry is also deeply personal. He invests in the spiritual formation of his children, Jaxon and Lorelei, creating devotional materials designed to help the next generation grow in faith. His passion is to equip believers of all ages to recognize the enemy's strategies, resist his assaults, and stand firm in the victory Christ has already won.

Through his unique background in both modern warfare and biblical teaching, Shane offers a field manual for spiritual combat—practical, strategic, and deeply rooted in Scripture. His mission is clear: to awaken the church to the reality of spiritual warfare and to raise up resilient, prepared soldiers of Christ who will fight faithfully until the day of ultimate victory.